Biomass Pelletization

Dr Jose Antonio Pérez Jiménez received his PhD from the University of Jaen (Spain) while he was working as a researcher and development biomass project at CIEMAT (Research Center for Energy, Environment, and Technology) He then worked at the University of Cadiz as a researcher and subsequently was the person responsible for the implementation of the R&D department of a private company focused on the development of renewable energies. Currently, he is the Head of Department of Biomass at CTAER (Advanced Technology Centre for Renewable Energies), where he collaborates with relevant international research groups working on biofuels and renewable energy production and has authored several publications with high international impact.

Dr Angela Garcia-Maraver received her PhD from the University of Granada (Spain), where she is developing her career as a researcher and lecturer in topics related to waste management and biomass energy at the Department of Civil Engineering. She has stayed as a researcher in internationally renowned Universities and has participated as a lecturer and consultant in public and private institutions. She has authored several publications about biomass and pelletization in international journals with high impact, as well as has participated in more than ten international conferences on renewable energies in the last 4 years .

Biomass Pelletization

Standards and Production

Edited by

A. Garcia-Maraver
University of Granada, Spain

&

J.A. Pérez-Jiménez
Advanced Technology Centre for Renewable Energies, Spain

WIT*PRESS* Southampton, Boston

Editors:

A. Garcia-Maraver
University of Granada, Spain

J.A. Pérez-Jiménez
Advanced Technology Centre for Renewable Energies, Spain

Published by

WIT Press
Ashurst Lodge, Ashurst, Southampton, SO40 7AA, UK
Tel: 44 (0) 238 029 3223; Fax: 44 (0) 238 029 2853
E-Mail: witpress@witpress.com
http://www.witpress.com

For USA, Canada and Mexico

Computational Mechanics International
25 Bridge Street, Billerica, MA 01821, USA
Tel: 978 667 5841; Fax: 978 667 7582
E-Mail: infousa@witpress.com
http://www.witpress.com

British Library Cataloguing-in-Publication Data

A Catalogue record for this book is available
from the British Library

ISBN: 978-1-78466-062-8
eISBN: 978-1-78466-063-5

Library of Congress Catalog Card Number: 2014958945

Printed by Lightning Source, UK.

Contents

CHAPTER 7

CHAPTER 8

CHAPTER 9

PREFACE

A substantial percentage of the world's demand for energy has been satisfied by the combustion of fossil fuels. During the past few years, it has been demonstrated that the continued reliance on fossil fuels as a major source of energy can have serious negative environmental implications. Using renewable sources of energy can curb these negative environmental issues. Up to now, two of the most common renewable sources of energy have been: biomass and hydroelectricity.

Biomass has been used with little or minimum processing to generate heat and electricity for a relatively long time. Biomass can also be treated chemically and biochemically to convert it into fuels. In addition to the relatively large numbers of hectares of species specifically planted as sources of biomass for energy production, a considerable amount of residues from some activities such as forestry and agricultural activities can also be used as sources of energy. One of the major drawbacks in the use of some of these residues is that the generation of these waste materials often takes place over relatively large areas. Consequently, the costs associated with collection and transportation of the biomass makes it impractical to use. One excellent option to reduce the costs associated with transport and use the residues as fuels is by means of densification or pelletization. Pelletization, not only improves the possibility of transporting the materials at an acceptable cost but also, if properly carried out, pelletization can improve the quality of the fuel and of the combustion process.

The authors of this book provide much needed detailed scientific information on the various aspects of pelletization, on the use of the material in combustion units, as well as on the emissions from the combustion of pellets.

The application of the concepts provided in this book would allow making a link between two very important global concerns: waste management and energy dependency.

Dr. Luis F. Diaz
President, CalRecovery, Inc.
USA

CHAPTER 1

Background

Angela Garcia-Maraver[1], Jose Antonio Perez-Jimenez[2] & Montserrat Zamorano[1]
[1]*Department of Civil Engineering, University of Granada, Campus de Fuentenueva, Granada, Spain.*
[2]*Department of Biomass, CTAER Andalucía Foundation, Scientific and Technology Park, Jaen, Spain.*

Abstract

Environmental and energy dependency problems stemming from high fossil fuel consumption have made necessary to develop new energy models. They must entail renewable and sustainable systems that are also efficient and cost-effective, in order to ensure safe energy supply. Among the renewable resources, biomass holds great energy potential; it also has a high environmental value because of its contribution to the development of sustainable systems and the reduction of the greenhouse effect. Pelletization is the most common mass and energy densification process in the realm of biomass solid fuels. Biomass pellets represent a particularly promising alternative to fossil fuels for heat and power production in the residential sector, where automated heating systems can be implemented. The growing domestic and industrial demand of biomass for heat and power production has led to a flourishing global pellet market in the past two decades, and continued growth is predicted for the coming years. Biomass in general – and pellets in particular – can thus be viewed as an alternative source deserving more research, dedicated to enhancing the quality of solid biofuels and the technologies applied to produce energy from biomass.
Keywords: Renewable energy, biomass, pelletization.

1 Introduction

The importance of energy in the generation of wealth and in economic development is universally acknowledged. Historic data demonstrate a strong relationship between energy availability and economic activity [1]. Yet economic development, industrialization and the implantation of economic models that base growth on the sustained increase of consumption have resulted in a society highly dependent on energy. The need for energy resources is increasing [2] as a direct result of the world's growing population and intense energy consumption linked to development and industrialization.

Nowadays, the vast majority of our energy is produced and consumed to a degree that is not feasible if technology remains constant and energy consumption keeps rising [3]. New energy models must be renewable and sustainable, efficient and economically effective, practical and safe [3,4]. Therefore, energy planning faces a historic challenge marked by the need to guarantee a qualified, safe and sufficient energy supply entailing no environmental imbalances. Until now, energy planning studies focused on forecasting the energy demand to be met, from a standpoint that views energy as a finite resource. The evident failure of this approach and the environmental impact of current energy trends, contributing heavily to global warming, urge that new energy models be put forth.

The growing awareness of fossil fuel consumption and environmental degradation have resulted in a generalized interest in renewable energy as a response to concerns about climate change [5,6] and energy efficiency, transmission, distribution and use [3]. Renewable energy resources are seen as the most efficient and effective road towards sustainable energy development and the prevention of environmental pollution.

In short, within the new energy framework, renewable energies are intended to cover part of the future fossil fuels demand [2] and therefore enable the world to reduce greenhouse emissions and make it less dependent on imported energy. Boosting the renewables industry will encourage technological innovation and employment.

Biomass can play an important role in this changing scenario, since it is regarded as a renewable and environmentally safe way of providing energy for electricity, heating and cooling [7]. As an abundant renewable energetic resource, biomass moreover entails significant socioeconomic and environmental benefits from the utilization of residues. By converting residue into energy, we re-appraise waste materials, mitigate the environmental impacts of waste disposal and reduce the mass and volume of waste.

2 Energy from Biomass

Biomass is the progenitor of fossil fuels, in the sense that natural gas, oil and coal are biomasses that were converted into concentrated energy forms by geological processes akin to pyrolysis (heating in the absence of oxygen) under pressure. Accordingly, with appropriate industrial processing, newly harvested biomass can be converted into the equivalent of current gas, liquid or solid fossil fuel resources.

Until the mid-19th century, biomass dominated global energy consumption. The steep increase in fossil fuel usage has resulted in a decline of biomass consumption for energy purposes over the last 50 years, however. Even so, biomass still provides about 14% of the world's annual energy consumption [8–10] and stands as the fourth largest energy resource in the world, behind coal, oil and natural gas [11].

The use of biomass as a traditional energy source for economically developing countries can play a pivotal role in helping the industrialized world to reduce the environmental impact of burning fossil fuels to produce energy [12].

Wood fuels, agricultural wastes and grasses are the most prominent biomass energy sources. When properly managed, biomass offers many benefits as a renewable and sustainable energy feedstock. It can significantly reduce net carbon emissions when compared with fossil fuels, making it a clean mechanism for reducing greenhouse gas emissions [13]. The cheapest biomass sources are the waste products from wood or agro-processing operations, but their supply is limited. To overcome this limitation, countries around the world are planting biomass crops for energy purposes.

Most developed and industrialized nations have begun developing technologies to use biomass more efficiently. In the United States and most of Europe, biomass has already penetrated the energy market. The future of biomass for energy production is promising.

2.1 Biomass definition and classification

2.1.1 Biomass definition

The European Technical Specification [14] defines biomass as 'all biological materials excluding those that were included in geological formations suffering a mineralization process.' Hence, coal, petroleum and natural gas, whose formation goes back to millions of years ago, are not comparable with the 'carbon neutral balance' of biomass.

A key distinction lies in the carbon cycle and carbon neutrality of biomass as opposed to those of the carbon in fossil fuels. Fossil fuels contain carbon that has been out of the atmosphere for millions of years, and when they are burned, that carbon returns to the atmosphere. It joins the carbon cycling between the atmosphere and the earth, causing the amounts of CO_2 in the atmosphere to increase and contribute to the greenhouse effect. Indeed, the primary source of increased CO_2 in the atmosphere since pre-industrial times is the combustion of fossil fuels [15].

In this context, the term 'carbon neutrality' is best understood as an attribute of biomass carbon reflecting the fact that it was only recently removed from the atmosphere and is part of a natural cycle.

Through photosynthesis, the carbon in atmospheric CO_2 becomes carbon in plant tissue, also called biomass. When biomass is burned, is decayed or is otherwise oxidized, the chemical energy is released and the CO_2 is put back into the atmosphere, completing a natural carbon cycle. If this cycle is in balance, it has a net zero impact on atmospheric carbon – a 'carbon neutral balance' [6,12,16,17]. This is an important distinction between biomass carbon and the carbon in fossil fuels.

The cycle is never truly in exact balance, however. For instance, if plants are removing carbon from the atmosphere faster than it is being returned, the cycle accomplishes net removal of carbon from the atmosphere, and stocks of stored carbon (primarily in forests) increase. On the other hand, if biomass carbon is being returned to the atmosphere faster than it is being removed by plants, the cycle is adding carbon to the atmosphere while stocks of stored carbon are decreasing [17]. Clearly, only sustainable management can lead to indefinitely renewable fuel supply.

2.1.2 Biomass classification

Biomass resources include natural and derived materials mainly classified in view of their origin [11,18–22].

Biomass sources can be divided into three principal groups: (1) natural biomass, directly available in natural ecosystems; (2) residual biomass, coming from the development of different activities; and (3) energy crops, with the sole objective of producing biomass for energy purposes. Residual biomass and energy crops are the ones most widely used to produce energy.

The product derived from a particular feedstock depends on the part of the plant and the conversion process used. Products that can be derived from biomass include solid biofuels, biodiesel, ethanol, butanol, methane, hydrocarbons and natural oils, which can be further processed into any number of desirable fuels [23].

Regarding solid biofuels, the European Committee for Standardization (CEN) under committee TC335 published 27 technical specifications (pre-standards) for solid biofuels during the period 2003–2006. These technical specifications have been upgraded to full European standards (EN).

The two most important technical specifications deal with classification and specification [24] and quality assurance for solid biofuels [25]. The classification of solid biofuels is based on their origin and source. The whole chain of fuel production chain can be unambiguously retraced.

Solid biofuels are divided into the following sub-categories for classification in EN 14961-1 [24]:

(a) Woody biomass: From trees, bushes and shrubs [24].
(b) Herbaceous biomass: From plants that have a non-woody stem and that die at the end of the growing season; this includes grains and their by-products such as cereals [24].
(c) Fruit biomass: From the parts of a plant, which hold seeds. If appropriate, the actual species (e.g. spruce, wheat) of biomass should be stated.
(d) Blends and mixtures: Blends are intentionally mixed biofuels, whereas mixtures are unintentionally mixed biofuels [24].

For the sake of clarity, demolition wood is not included in the scope of the EN 14961-1. Demolition wood is 'used wood arising from the demolition of buildings or civil engineering installations' [14].

2.2 Biomass advantages and barriers

2.2.1 Biomass advantages

Increases in the price of fossil fuels, growing environmental concerns regarding their use and impact (including climate change) and considerations regarding the security of energy supply have motivated the rising use of renewable energy sources worldwide [26].

Biomass-to-energy facilities present many benefits. Biomass is a renewable resource, and generating electricity and other energy products from biomass offsets consumption of fossil fuels. Biomass is typically waste material from another industry, such as logging or dairy operations, and converting it to energy not only reduces disposal but also mitigates the environmental impact that such waste can have on air, groundwater and surface water quality. These advantages, together with those described below, make biomass a very important element of territorial balance, especially in rural areas.

(a) *Reduction of pollutant emissions such as CO, HC and* NO_x: Using biomass for energy production means reduced CO_2 emissions [27], reduced SO_2 formation through a decrease in fuel-bound sulfur and reduced NO_x formation through a reduction in fuel-bound nitrogen [28–30]. The low levels of particulates, CO and C_xH_y, contribute to maintaining the photochemistry of the atmosphere [31].
(b) *Biomass carbon neutrality, no contribution to the greenhouse effect*: One of the most important advantages related to the use of biomass is the carbon neutrality cycle [32–34] previously described. An increase of 1695 MW in electric power generation from biomass and 582.5 ktoe in primary energy from thermal biomass energy would avoid the emission of nine million tonnes of CO_2 to the atmosphere [35].
(c) *Reduction of risks involving forest fires and insect plagues*: The exploitation of agricultural biomass as a fuel for biomass boilers is one means of keeping forests clean.
(d) *Exploitation of agricultural wastes*: Agricultural cuttings and prunings, traditionally burnt in the country, can find a new market as residual biomass to produce energy. In this sense, the use of residual biomass improves waste management and mitigates the environmental impact of waste disposal, according to the Directive [36] on waste.
(e) *Use of fallow lands for energy crops*: Reforestation with energy crops can increase water retention as well as reduce the degradation and erosion of the soil. When correctly managed, this energy resource stops erosion and contamination of the water, because the vegetation cover reduces superficial run-off and stabilizes particles in the soil, preserving animal and vegetal biodiversity.
(f) *Reduced dependence on foreign oil, given that biofuels are the only renewable liquid fuels currently available*: Biomass and biological wastes are autonomous fuel resources, partially averting dependence on foreign energy supplies, improving a country's trade balance and economic sustainability [37,38].
(g) *Socioeconomic improvement of rural areas*: Promoting biomass production for energetic purposes can lead to new activity in rural areas, which may become energetically self-sufficient by using local biomass. The development of biomass and energy crops stimulates the job market, avoiding the depopulation of rural areas while supporting local agriculture. In this way, the impact on food production can be minimized [39].

2.2.2 Biomass barriers

In contrast to the benefits, there are significant barriers to biomass-to-energy facilities.

Compared with conventional heating systems, such as oil or gas boilers, biomass technology still entails disadvantages in terms of space, efficiency, particulate emissions and maintenance. In light of the low energy densities yielded, the collection and transportation of biomass can be cost prohibitive. On the other hand, biomass fuels tend to have a high moisture content, which adds weight and increases the cost of transportation. The moisture content also decreases combustion performance. Nevertheless, an intelligent design and sophisticated technology can minimize these disadvantages [40].

In addition to the logistic barriers regarding the use of biomass, it is very important to use it in a sustainable way. This calls for comprehensive management of natural resources such as land and water. Unsustainable biomass production would nullify the climate-related environmental advantage of bioenergy. It is necessary to reliably demonstrate that the advantages of biofuels exceed the cost of the potential environmental damage caused by their production. This stands as a major challenge in the context of biomass and biofuel production. Sound criteria to ensure the sustainable production of biomass are urgently needed [11].

2.3 Biomass energy conversion technologies

There are many types of biomass (organic matter such as plants, residues from agriculture and forestry, and the organic component of municipal and industrial wastes) that can be used to produce fuels, chemicals and power. This flexibility has given rise to diverse biomass technologies (Fig. 1).

Factors that influence the choice of the conversion process are: the type and quantity of biomass feedstock; the desired form of the energy, i.e. end-use requirements; environmental standards; economic conditions; and project-specific factors. In many situations, one particular form will determine the processing route, or the available types and quantities of biomass will be determinant [12].

2.4 Biomass potential and consumption

Biomass is the fourth largest energy source after coal, oil and natural gas [11], and consequently is the largest and most important renewable energy option at present.

It is, together with the other renewable energy options, capable of providing all the energy services required in a modern society, both locally and in most parts of the world. Renewability and versatility are important advantages of biomass as an energy source. Compared with other renewables, biomass resources are common all over the world [42,43]. Therefore, this widespread resource for energy production holds great potential.

2.4.1 Global biomass potential and consumption

Biomass is the primary source of energy for nearly 50% of the world's population [44], and wood biomass is a major renewable energy source in the developing

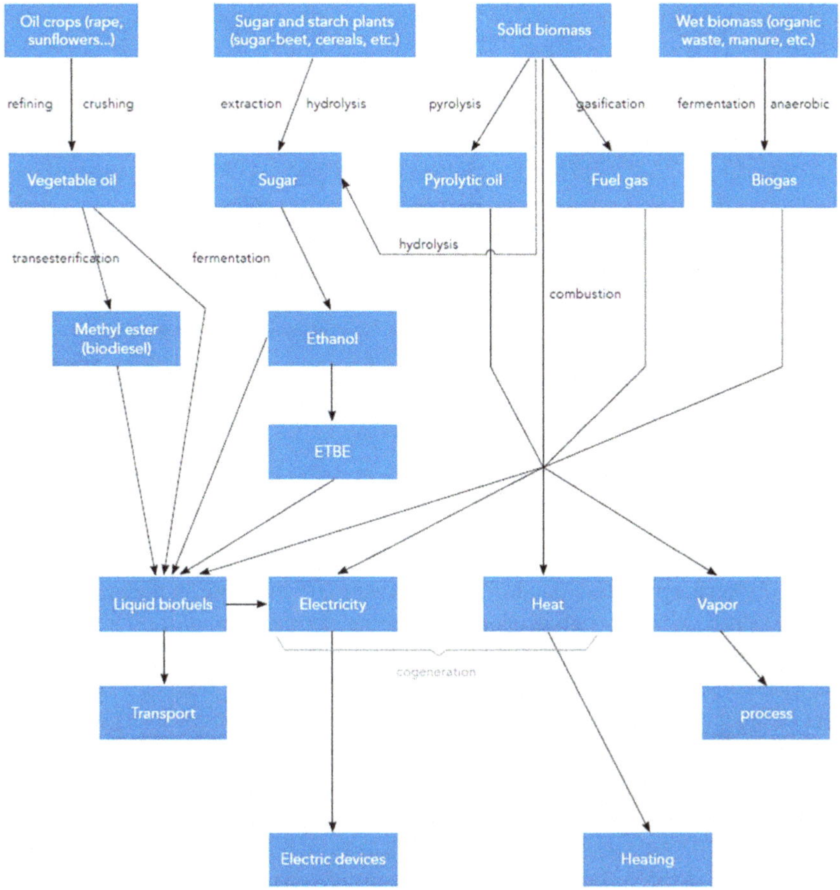

Figure 1: Biomass energy conversion overview [41].

world, representing a significant proportion of the rural energy supply [45]. Over the past decade, the number of countries exploiting biomass opportunities for the provision of energy has increased rapidly and has helped to make biomass an attractive and promising option in comparison to other renewable energy sources. The global use of biomass for energy production increases continuously and has doubled in the last 40 years [46].

The annual global primary production of biomass is equivalent to the 4,500 EJ of solar energy captured each year [47]. The potential of global biomass as a sustainable energy source is widely recognized. At present, a bioenergy supply of 270 EJ – possible to achieve on a sustainable basis using only 6% of the annual global primary production of biomass – could cover almost 50% of the world's total primary energy demand [11]. However, the contribution of biomass to the global energy demand of 532 EJ in 2010 amounted to just 10% [48].

The future potential for energy from biomass depends in part on land availability. The amount of land currently devoted to growing energy crops for biomass fuels is

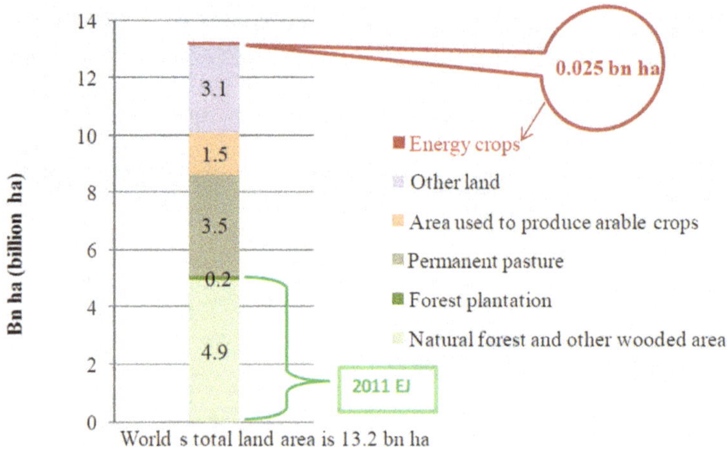

Figure 2: Distribution of land use types in world total land area [49,50].

only 0.19% of the world's total land area (0.025 billion hectares) and only 0.5%–1.7% of global agricultural land (Fig. 2). A mere 10% increase in the efficiency of biomass production through irrigation, manuring, fertilizing and/or improved management, e.g. through the cultivation of idle land, would create energy equivalent to the total current global energy demand [11]. Standing forest stock is a large reservoir of bioenergy, in line with the theoretical potential of biomass energy.

There are many factors signaling the potential of biomass, and numerous studies have attempted to estimate the future demand and supply of bioenergy [43,51–55]. Published estimates of the total global bioenergy production potential in 2050 ranged from 33 to 1,135 EJ annually [56], 0 to 358 EJ annually coming from woody biomass [56,57]. Energy crops from surplus agricultural land have the largest potential contribution, i.e. 0–988 EJ/year [56].

The vast range of figures can be attributed to: differences in the type of biomass included; differences in the theoretical, technical, economic or ecologic limitations related to the supply of woody biomass for energy use; differences in data on key parameters, such as the consumption of wood fuel, the annual growth of forests and the efficiency of conversion; and differences in scope, whereby most of the existing bioenergy potential assessments focused on either the demand [58] or the supply [59] of bioenergy and consequently ignored demand–supply interactions [60]. Moreover, as Smeets and Faaij [60] pointed out, most studies reviewed ignored existing studies on the demand and supply of wood [61,62], despite the extensive literature and data on the subject [63,64]. The diverse interpretations stem from large differences in demand and energy mix, as a result of variations in population dynamics, and economic and technological development.

The global primary energy demand was 302 EJ in 1980, 420 EJ in 2000, 470 EJ in 2007 [65] and 532 EJ in 2010 [48]; yet, it was expected to increase to 591 and 712 EJ, respectively, by 2015 and 2030 [66], and to 826 EJ (as averaged for different scenarios) by 2050 [49]. However, even in the case featuring the highest consumption scenarios of 1,041 EJ in 2050 [49], the 1,342 EJ which is the average of the cited bioenergy production potentials in 2050 would suffice to meet this demand [11]. The current theoretical potential of biomass energy is 2,900 EJ [67]. The current stock of standing forest, with a total energy content corresponding to 2,011 EJ is a large reservoir of bioenergy (Fig. 2). Overall, the world's bioenergy potential appears to be large enough to meet global energy demand in 2050.

2.4.2 European biomass potential and consumption

Europe is in the midst of a dramatic energy transition, away from fossil fuels, and in some cases also away from nuclear power, to renewable energy. Bioenergy is the major renewable energy source, accounting for almost 70% of European renewables [68,69], and showing steady growth. This development did not start with the 20–20–20 goals formulated in 2007, or the Renewable Energy Directive adopted in 2009. Before these goals were in place, renewable energy sources had already almost doubled their production with respect to the early 90s [69].

The national action plans presented by the EU member states will further stimulate this development, particularly for the 'slow starters'. One requirement for success, however, is for governments to come forward with strong incentives, not only for power production and the transport sector, but in particular for the heating sector. Paying extra for carbon dioxide emissions, in other words, or instating a common European carbon tax, would give the push needed to convert heating systems from fossil fuels to biomass [70–73].

The consumption of renewable energies has significantly increased in recent years. While hydropower stagnates, solar and wind energy show impressive growth rates. Yet starting from a relatively small market share, biomass is by far the most important source of renewable energy in Europe. Biomass represented 68.6% of the consumption of renewable energies in the EU in 2009, remaining stable with respect to the previous year (Table 1).

At the EU level, in absolute terms, the total gross final energy consumption fell from 1193.3 Mtoe in 2006 to 1113.6 Mtoe in 2009, while consumption of renewable energy rose from 129.1 Mtoe in 2006 to 152.6 Mtoe in 2009, with an average biomass share of 65% [69].

Figure 3 depicts the bioenergy balance for 2008. Over one-third of the biomass consumed in EU is fed to electricity, combined heat and power (CHP) and district heating plants, while the rest is consumed in the private, commercial and industrial sectors for heating purposes. About 10% is used as biofuels in the transport sector, and 60% of biomass from CHP plants is used as recovered heat.

The Renewable Heating and Cooling (RHC) Platform expects biomass use to more than double by 2020, and to reach around 370 Mtoe of primary energy in 2050 [74], mostly to meet the heat demand (231 Mtoe total contribution to heat demand in 2050). By 2020, the biomass supply should have increased significantly to meet the demand of all sectors of heat, electricity and transport biofuels. Such developments could be linked to changes in other sectors, including agriculture

Table 1: Gross inland consumption of renewables 1995–2009 in EU27 (stacked, Mtoe) [69].

Year	Renewables	Biomass and waste	Solar	Geothermal	Hydro power	Wind energy	Biomass share (%)
2000	98.2	59.5	0.4	3.4	30.4	1.9	60.6
2001	101.4	60.4	0.5	3.6	32.0	2.3	59.6
2002	99.8	62.1	0.5	3.9	27.1	3.0	62.2
2003	107.8	67.8	0.6	5.3	26.3	3.8	62.9
2004	116.2	72.5	0.7	5.3	27.8	5.0	62.4
2005	120.9	77.5	0.8	5.3	26.4	6.0	64.1
2006	129.1	83.5	1.0	5.5	26.5	7.0	64.7
2007	143.1	91.8	1.2	5.7	26.6	8.9	64.2
2008	144.2	98.2	1.7	5.7	28.1	10.2	68.1
2009	152.6	104.7	2.4	5.8	28.1	11.4	68.6

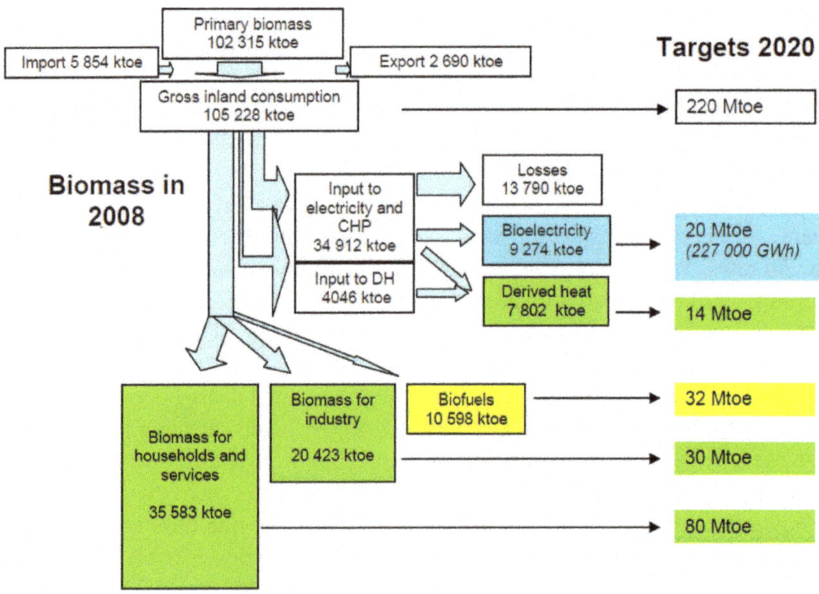

Figure 3: European bioenergy balance in 2008 and 2020 targets [69,74].

and materials production. The greatest contribution to the biomass supply comes from energy crops, by-products from agriculture and forest logging residues.

To better discern the future role of bioenergy in Europe, the potential of biomass can be envisaged in three particularly relevant sectors: agriculture, forestry and waste. They involve categories of biomass production such as biofuel crops,

Table 2: Environmentally compatible bioenergy potential (Mtoe) in the EU25 [68].

	2003	2010	2020	2030
Total	69	187	228	284
Wood direct from forest	–	43	43	55
Wastes and residues	67	100	100	102
Energy crops from agriculture	2	44	85	122

agricultural by-products or primary and secondary forestry residues. The forest (41%) and waste (38%) sectors contribute the lion's share of potential. The remaining 21% could come from the agricultural sector and be scattered over many different small categories [69]. Within the agricultural group, the largest contribution will most likely come from straw, dedicated cropping and prunings.

The environmentally compatible potential (Table 2) designates the quantity of biomass that is technically available for energy generation based on the assumption that this places no additional pressures on biodiversity, soil and water resources when compared with development without increased bioenergy production. Moreover, it should be in line with other current and future environmental policies and objectives. This concept is tied to a number of environmental constraints for calculating the available technical potential; and it was assumed that greenhouse gas emissions would be further reduced by 20% in 2020 and 40% in 2030 by the EU-25 [68]. The data in Table 2 show the production potential for biomass in the EU 25 as calculated by the European Environmental Agency [68]. The figures for EU 27 can be estimated as 10% higher. The main growth is seen in the sector of 'waste and residues' and 'energy crops from agriculture', due to the Utilized Agricultural Area (UAA), which represents 38% of the whole EU-27 territorial area [69].

The agricultural biomass resource for bioenergy depends mainly on the available land area and the yield of bioenergy crops grown. Priority was given to domestic food production needed to satisfy food demand. In consequence, the share of UAA in the total area varies greatly from country to country, from just 2% in Latvia or 7% in Finland and Sweden, to 65% in the United Kingdom. As part of the UAA, arable land represents almost one quarter of the whole EU-27 territory and Denmark has the highest share of arable land (57%). Permanent grassland represents 14% of EU-27 territory; whereas more than 45% of the land in Ireland and the United Kingdom is used for permanent grassland, extreme northern and southern countries (Finland and Cyprus) have less than 1% of their land under permanent grassland. Finally, land under permanent crops represents <3% in the EU-27. Yet, several southern European countries have a higher share of land under permanent crops: 10% in Spain, 9% in Greece and Italy, and 8% in Portugal [69].

3 Densification of Biomass

Because of its high moisture content, irregular shape and size, and low bulk density, biomass is very difficult to handle, transport, store and utilize in its original form [75,76]. To make the biomass available for a variety of applications, certain challenges with the use of biomass materials must be resolved.

The densification of biomass materials could reduce the costs of transportation, handling and storage. Having uniform shape and size, densified products can be easily handled and stored using standard equipment, and easily adopted in direct combustion or co-firing systems with coal, gasification, pyrolysis or other biomass-based conversions [18,76].

Conventional processes for pressure-assisted densification can be classified into three types: extruding, pelleting and briquetting [75,77].

In an extruder, the raw material is compressed by a screw or a piston through a die to form compacted particles, often of cylindrical shape. The diameter of extrusion products may range from 20 to 100 mm. Extrusion technologies are commonly used for biomass densification in Asian countries such as India, Japan and Thailand [78–81].

With pelleting equipment, the feed material is pressed through open-ended cylindrical holes (dies) made in the shape of a ring. One to three small rotating rolls push the feed material into the holes of the die, from the inside towards the outside of the ring. The skin friction between the feed particles and the wall of the die resists the free flow of feed, and thus the particles are compressed against each other inside the die to form pellets. One or two adjustable knives placed outside the ring cut the pellets into desired lengths. The diameter of the pellets may range from 4.8 to 19.0 mm, and the length of the pellets may range from 12.7 to 25.4 mm [75]. Pelleting technology is commonly used in the United States and in Europe for producing animal feed.

In a roll press briquetter/compactor, the material is densified by compression between two counter-rotating rolls. Initial densification of the material may involve compressing the material with a tapered auger in the feed mechanism. The primary purpose of this initial densification phase is to remove air from low bulk density material. The final compaction occurs because of high pressures created as the material flows between the two rollers. The roll surfaces have pockets to form briquettes of the desired size and shape when the material passes between the rolls. These briquettes are easily separated and handled after leaving the machine. The densified products are most commonly pillow-shaped, with a size of 10–40 mm or larger [75]. The roll press is generally used for densifying coal, minerals and metals in the United States and in Europe [77].

Pelletization is the most common densification process used for solid fuel applications. Wood pellets from forestry residues are already well-established products in some European countries. However, in other countries this technique is still not widely used. For a successful end product, the quality of the densification process must meet consumer requirements and market standards. Because the final quality of pellets will vary depending on the raw material properties and the manufacturing process [82], these variables should be controlled to optimize production efficiency and improve the quality of finished product.

3.1 Pelletization advantages

Densified fuel pellets afford several advantages. The higher bulk and energy densities result, respectively, in lower transport costs and higher energy conversion efficiency, while the reduced moisture content increases the long-term storage capability [83]. As detailed below, such advantages are likely to stimulate the use of pellets in small stoves and boilers, where the enhanced properties will outweigh the extra cost involved [84]:

(a) The bulk density of biomass can be increased from an initial bulk density (including baled density) of 40–200 kg/m^3 to a final bulk density of 600–800 kg/m^3 [85].
(b) The uniform size and shape make pelleted biomass easier to handle using existing handling and storage equipment; and by increasing its durability, damage to the material can be avoided. Therefore, the densification process reduces transportation costs and provides for better handling and feeding of the biomass with less dust formation [76,86], reducing cleaning and maintenance activities and making pellets an optimum fuel for domestic heating systems.
(c) High energetic density, depending on the raw material, can be achieved through compaction and low moisture content (6%–8%). In most cases, the high calorific value ranges from 16.5 to 19 MJ/kg.
(d) The versatility of pellets stems from the possibility of producing them from a wide range of raw materials. The main ones are wood residues – sawdust, wood shavings and wood chips – agro-residues such as straw, and waste products of the food industry and fuel crops [87,88].

3.2 Pelletization process

Before being fed into any densification equipment, the feed material is ground to a desired particle size, subjected to pre-treatments such as mixing with additives or fats, and conditioned with steam or expander to increase the temperature and/or moisture level. During the pelletization process, the raw material is dried, ground, homogenized and densified (Fig. 4), improving its storage, transport and handling characteristics [89].

The principal stages of the process are as follows:

(a) *Size reduction*: Whether the raw material is wood, grass, straw or any other type of biomass, it must be reduced to a sufficiently uniform size for the pellet mill. The general rule is that the milled material going into the pellet mill must be smaller than the die holes in the pellet mill to avoid blocking the holes [90]. For example, to produce 6-mm pellets the milled material must be smaller than 6 mm.

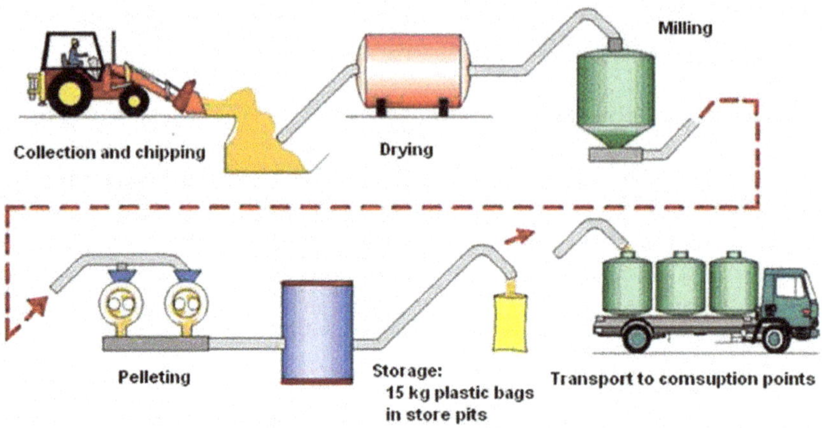

Figure 4: Diagram of the pelletization process [92].

There are various types of equipment to conduct size reduction (chippers/shredders, hammer mills); each has different abilities and strengths [91]. In some cases, one piece of equipment is required, but the most common process involves two different stages of grinding: a chipper before the dryer and a hammer mill after the dryer, capable of reducing the particle size to powder.

(b) *Drying*: The moisture content of the raw material is an important parameter for the pelletizing (compaction) process because moisture facilitates heat transfer and friction. If the moisture content is too low, the friction between the raw material and the die will become too high, resulting in increasing power consumption, and eventually the die holes might become clogged. The moisture content is also important for the thermal softening and self-bonding of individual particles in the pellet [93].

To produce quality pellets, the moisture percentage of the raw material therefore must be homogeneous as well as adequate for the pelletization process. Most materials produce the highest quality pellets with a moisture content of around 10%–15% (raw weight); however, the moisture content to produce high-quality pellets is specific to each raw material [91].

Since it is not possible to reach such low moisture contents by natural drying, the material has to be artificially dried. This is usually done using rotary/drum dryers.

(c) *Pelletization*: The friction between the woody material and the die during the forced passage results in a rise in temperature. This, in combination with the moisture and lignin contents of the raw material, softens the wood components and is assumed to have a crucial influence on the self-bonding processes [86,93]. This means that it is not necessary to add binders for most biomass pellets, the composition of lignocellulosic materials working as a binder once the temperature inside the pellet press has increased.

There are two types of pellet mills: flat die and round die ones. Generally, flat die pellet mills are used for small to medium scale pellet production, whereas round die pellet mills are used for medium to large-scale production. The flat die pellet mill works on the principle of material falling from above onto the rollers, rotating over the pellet mill die. The round die pellet mill comprises a vertical ring die with rollers on the inside, applying pressure against ring die [91].

(d) *Cooling, storing and packaging*: The temperature of the products leaving the densification equipment is generally higher than the conditioned feed due to frictional heat developed in the pressing systems, for which reason the densified products are cooled using ambient air before storing/packaging. This final step also results in a reduction of moisture content, and the pellet hardens [28].

Once the pellets are produced and cooled, they are stored in silos before being transported by tankers to the consumer, or packed in bags of 15 kg.

References

[1] Spalding-Fecher, R., Winkler, H. & Mwakasonda, S., Energy and the World Summit on Sustainable Development: what next? *Energy Policy*, **33**, pp. 99–112, 2005.

[2] Marcos Martín, F., Certificación AE de N y. *Biocombustibles sólidos de origen forestal*, Madrid: Aenor, 2001.

[3] Kaygusuz, K. & Sari, A., Renewable energy potential and utilization in Turkey. *Energy Convers Manag*, **44**, pp. 459–478, 2003.

[4] Chum, H.L. & Overend, R.P., Biomass and renewable fuels. *Fuel Process Technol*, **71**, pp. 187–195, 2001.

[5] Akella, A.K., Sharma, M.P. & Saini, R.P., Optimum utilization of renewable energy sources in a remote area. *Renew Sustain Energy Rev*, **11**, pp. 894–908, 2007.

[6] Cowie, A.L. & David Gardner, W., Competition for the biomass resource: greenhouse impacts and implications for renewable energy incentive schemes. *Biomass and Bioenergy*, **31**, pp. 601–607, 2007.

[7] Fernández Llorente, M.J. & Carrasco García, J.E., Comparing methods for predicting the sintering of biomass ash in combustion. *Fuel*, **84**, pp. 1893–1900, 2005.

[8] Werther, J., Saenger, M., Hartge, E.-U., Ogada, T. & Siagi, Z., Combustion of agricultural residues. *Prog Energy Combust Sci*, **26**, pp. 1–27, 2000.

[9] Purohit, P., Tripathi, A.K. & Kandpal, T.C., Energetics of coal substitution by briquettes of agricultural residues. *Energy*, **31**, pp. 1321–1331, 2006.

[10] Zeng, X., Ma, Y. & Ma, L., Utilization of straw in biomass energy in China. *Renew Sustain Energy Rev*, **11**, pp. 976–987, 2007.

[11] Ladanai, S. & Vinterbäck, J., Global potential of sustainable biomass for energy. SLU Report. ISSN 1654-9406, 2009.

[12] McKendry, P., Energy production from biomass (part 1): overview of biomass. *Rev Issue*, **83**, pp. 37–46, 2002.
[13] Junfeng, L. & Runqing, H., Sustainable biomass production for energy in China. *Biomass and Bioenergy*, **25**, pp. 483–499, 2003.
[14] EN14588. Solid biofuels. Terminology, definitions and descriptions, 2011.
[15] Denman, K.L., Brasseur, G., Chidthaisong, A., Ciais, P., Cox, P.M., Dickinson, R.E., Hauglustaine, D., Heinze, C., Holland, E., Jacob, D., Lohmann, U., Ramachandran, S., da Silva Dias, P.L., Wofsy, S.C. & Zhang, X., Couplings between changes in the climate system and biogeochemistry. *Climate Change 2007: The Physical Science Basis*, Cambridge University Press: Cambridge, UK and New York, NY, p. 499, 2007.
[16] Varela, M., Lechón, Y. & Sáez, R., Environmental and socioeconomic aspects in the strategic analysis of a biomass power plant integration. *Biomass and Bioenergy*, **17**, pp. 405–413, 1999.
[17] Miner, R., Biomass Carbon Neutrality. 2010.
[18] Vinterbäck, J., Pellets 2002: the first world conference on pellets. *Biomass and Bioenergy*, **27**, pp. 513–520, 2004.
[19] VIEWLS, Shift Gear to Biofuels – Results and recommendations from the VIEWLS project, 2005.
[20] Junta de Andalucia (JA), Situación de la biomasa en andalucía. 2008.
[21] Rofiqul Islam, M., Rabiul Islam, M. & Rafiqul Alam Beg, M., Renewable energy resources and technologies practice in Bangladesh. *Renew Sustain Energy Rev*, **12**, pp. 299–343, 2008.
[22] Van Loo, S. & Koppejan, J., *The Handbook of Biomass Combustion and Co-firing*, Earthscan: Sterling, VA, 2008.
[23] Tumuluru, J.S., Wright, C.T., Boardman, R.D., Yancey, N.A. & Sokhansanj, S., *A Review on Biomass Classification and Composition, Co-firing Issues and Pretreatment Methods*, ASABE, 2011.
[24] EN14961-1. Solid biofuels. Fuel specification and classes. Part 1: general requirements, 2010.
[25] EN15234-1. Solid biofuels. Fuel quality assurance. Part 1: general requirements, 2011.
[26] Luque, R., Herrero-Davila, L., Campelo, J.M., Clark, J.H., Hidalgo, J.M., Luna, D., Marinas, J.M. & Romero, A.A., Biofuels: a technological perspective. *Energy Environ Sci*, **1**, pp. 542–564, 2008.
[27] Bram, S., De Ruyck, J. & Lavric, D., Using biomass: a system perturbation analysis. *Appl Energy*, **86**, pp. 194–201, 2009.
[28] Lehtikangas, P., Quality properties of pelletised sawdust, logging residues and bark. *Biomass and Bioenergy*, **20**, pp. 351–360, 2001.
[29] Li, Z., Zhao, W., Li, R., Wang, Z., Li, Y. & Zhao, G., Combustion characteristics and NO formation for biomass blends in a 35-ton-per-hour travelling grate utility boiler. *Bioresour Technol*, **100**, pp. 2278–2283, 2009.
[30] Qian, F.P., Chyang, C.S., Huang, K.S. & Tso, J., Combustion and NO emission of high nitrogen content biomass in a pilot-scale vortexing fluidized bed combustor. *Bioresour Technol*, **102**, pp. 1892–1898, 2011.

[31] García, R., Pizarro, C., Lavín, A.G. & Bueno, J.L., Characterization of Spanish biomass wastes for energy use. *Bioresour Technol*, **103**, pp. 249–258, 2012.
[32] Wahlund, B., Yan, J. & Westermark, M., Increasing biomass utilisation in energy systems: a comparative study of CO_2 reduction and cost for different bioenergy processing options. *Biomass and Bioenergy*, **26**, pp. 531–544, 2004.
[33] Gil, M.V., Casal, D., Pevida, C., Pis, J.J. & Rubiera, F., Thermal behaviour and kinetics of coal/biomass blends during co-combustion. *Bioresour Technol*, **101**, pp. 5601–5608, 2010.
[34] Gil, M.V., Oulego, P., Casal, M.D., Pevida, C., Pis, J.J. & Rubiera, F., Mechanical durability and combustion characteristics of pellets from biomass blends. *Bioresour Technol*, **101**, pp. 8859–8867, 2010.
[35] IDAE, Energía de la biomasa. 2007.
[36] D2008/98/EC. Waste Framework Directive 2008.
[37] Wolf, A., Vidlund, A. & Andersson, E., Energy-efficient pellet production in the forest industry – a study of obstacles and success factors. *Biomass and Bioenergy*, **30**, pp. 38–45, 2006.
[38] Demirbas, A., Biofuels sources, biofuel policy, biofuel economy and global biofuel projections. *Energy Convers Manag*, **49**, pp. 2106–2116, 2008.
[39] Bahng, M.-K, Mukarakate, C., Robichaud, D.J. & Nimlos, M.R., Current technologies for analysis of biomass thermochemical processing: a review. *Anal Chim Acta*, **651**, pp. 117–138, 2009.
[40] Fiedler, F., The state of the art of small-scale pellet-based heating systems and relevant regulations in Sweden, Austria and Germany. *Renew Sustain Energy Rev*, **8**, pp. 201–221, 2004.
[41] UNF, Sustainable Bioenergy Development in UEMOA Member Countries. 2008.
[42] Santamarta, J., Las energías renovables son el futuro. World Watch 2004.
[43] WEC, Survey of Energy Resources. 2004.
[44] Karekezi, S. & Kithyoma, W., Bioenergy and the poor. *Bioenergy Agric - Promises Challenges*, **14(11)**, 2006.
[45] Hashiramoto, O., Wood-product trade and policy issues. *Cross-Sectoral Policy Developments in Forestry*, eds Y.C. Dubé & F. Schmithüsen, ISBN 978-1, p. 24, 2007.
[46] World-Bank, World Development Indicators (WDI). WDI Online Databases 2009, available at http://web.
[47] Sims, R.E.H., *Bioenergy Options for a Cleaner Environment in Developed and Developing Countries*, Vol. 1, Elsevier: Amsterdam and Boston, 2004.
[48] IEA, Key World Energy Statistics 2012. 2012.
[49] Smeets, E., Faaij, A. & Lewandowski, I., *A Quick Scan of Global Bioenergy Potentials to 2050. An Analysis of the Regional Availability of Biomass Resources for Export in Relation to the Underlying Factors*, Vol. NWS-E; 200, Utrecht University: the Netherlands, 2004.
[50] Faaij, A., Sustainable Biofuels. *Presentation at Roclfeller Bellagio Conference on North-South Biopact, Bellagio, Italy,* 2008.

[51] Fischer, G. & Schrattenholzer, L., Global bioenergy potentials through 2050. *Biomass and Bioenergy*, **20**, pp. 151–159, 2001.
[52] Fujino, J., Yamaji, K. & Yamamoto, H., Biomass-Balance Table for evaluating bioenergy resources. *Appl Energy*, **63**, pp. 75–89, 1999.
[53] Hall, D.O., Rosillo-Calle, F., Williams, R.J. & Woods, J., *Biomass for Energy: Supply Prospects*, Earthscan: London, 1993.
[54] Lashof, D.A. & Tirpak, D.A., *Policy Options for Stabilising Global Climate*, Hemisphere Publishing Corporation, 1990.
[55] Rogner, H.H., Energy resources. *World Energy Assessment*, UNPD: Washington, DC, p. 135, 2000.
[56] Hoogwijk, M., Faaij, A., van den Broek, R., Berndes, G., Gielen, D. & Turkenburg, W., Exploration of the ranges of the global potential of biomass for energy. *Biomass and Bioenergy*, **25**, pp. 119–133, 2003.
[57] Sørensen, B., Long-term scenarios for global energy demand and supply: four global greenhouse mitigation scenarios. Energy & Environment Group, Roskilde University: Roskilde, Denmark, 1999.
[58] WEC, *New Renewable Energy Sources. A Guide to the Future*, World Energy Council/Kogan Page Ltd: London, UK, 1994.
[59] Yamamoto, H., Yamaji, K. & Fujino, J., Evaluation of bioenergy resources with a global land use and energy model formulated with SD technique. *Appl Energy*, **63**, pp. 101–113, 1999.
[60] Smeets, E. & Faaij, A. Bioenergy potentials from forestry in 2050 – an assessment of the drivers that determine the potentials. *Clim Change*, **8**, p. 353, 2007.
[61] Lazarus, M.L., Greber, L., Hall, J., Bartels, C., Bernow, S., Hansen, E., *et al. Towards a Fossil Free Energy Future: The Next Energy Transition. A Technical Analysis for Greenpeace International.* Boston Center: Boston, 1993.
[62] Sørensen, B., Biomass for energy: how much is there? *Hear. biofuels Transp.*, Vol. Danish Par, p. 149, 2001.
[63] Sedjo, R.A. & Lyon, K.S., *Timber Supply Model 96: A Global Timber Supply Model with a Pulpwood Component*, 1996.
[64] Solberg, B., Brooks, D., Pajuoja, H., Peck, T.J. & Wardle, P.A., *Long-Term Trends and Prospects in World Supply and Demand for Wood and Implications for Sustainable Forest Management*, Vol. 6. European Forest Institute: Joensuu, Finland, 1996.
[65] IEA, Key World Energy Statistics 2009. 2009.
[66] IEA, World Energy Outlook 2009. 2009.
[67] WEA, Energy and the Challenge of Sustainability. United Nations Development Programme, 2000.
[68] EEA, How much biomass can Europe use without harming the environment? 2005.
[69] AEBIOM, Annual Statistical Report on the contribution of biomass to the energy system in the EU27. 2011.
[70] EC, The share of renewable energy in the European Union. 2004.
[71] EREC, Renewable energy target for Europe-20% by 2020. 2004.

[72] EEA, Climate change and a European low-carbon energy system. 2005.
[73] Ragwitz, M., Scheich, J., Huber, C., Resch Faber, T.G., Voogt, M., Ruijgrok, W., *et al*. FORRES 2020-Analysis of the renewable sources′ evolution up to 2020. 2005.
[74] RHC, Common Vision for the Renewable Heating & Cooling sector in Europe. 2011.
[75] Li, Y. & Liu, H., High-pressure densification of wood residues to form an upgraded fuel. *Biomass and Bioenergy*, **19**, pp. 177–186, 2000.
[76] Kaliyan, N. & Vance Morey, R., Factors affecting strength and durability of densified biomass products. *Biomass and Bioenergy*, **33**, pp. 337–359, 2009.
[77] Pietsch, W., *Agglomeration Processes – Phenomena, Technologies, Equipment*, Weinheim: Wiley-VCH, 2002.
[78] Bhattacharya, S.C., Saunier, G.Y., Shah, N. & Islam, N., Densification of biomass residues in Asia. *Bioenergy 84. Proc. Conf., 15–21 June, 1984, Göteborg, Sweden. Vol. III. Biomass Convers.*, eds H. Egnéus & A. Ellegård, p. 559, 1984.
[79] Grover, P.D. & Mishra, S.K., Biomass briquetting: technology and practices. *Regional Wood Energy Development Program in Asia*, Vol. 46, Food and Agriculture Organization of the United Nations: Bangkok, Thailand, 1996.
[80] Varadharaju, N. & Gothandapani, L., Design and development of equipment for pelleting decomposed coir pith. *Agric Mech Asia, Africa Lat Am*, **29**, p. 33, 1998.
[81] Klass, D.L., Virgin biomass production. *Biomass for Renewable Energy, Fuels, and Chemicals*, ed. D.L. Klass, Academic Press: San Diego, pp. 91–136, 1998.
[82] Carone, M.T., Pantaleo, A. & Pellerano, A., Influence of process parameters and biomass characteristics on the durability of pellets from the pruning residues of Olea europaea L. *Biomass and Bioenergy*, **35**, pp. 402–410, 2011.
[83] Holm, J.K., Henriksen, U.B., Hustad, J.E. & Sorensen, L.H., Toward an understanding of controlling parameters in softwood and hardwood pellets production. *Energy Fuels*, **20**, pp. 2686–2694, 2006.
[84] Carroll, J. & Finnan, J., Emissions and efficiencies from the combustion of agricultural feedstock pellets using a small scale tilting grate boiler. *Biosyst Eng*, **115**, pp. 50–55, 2013.
[85] Obernberger, I. & Thek, G., Physical characterisation and chemical composition of densified biomass fuels with regard to their combustion behaviour. *Biomass and Bioenergy*, **27**, pp. 653–669, 2004.
[86] Gilbert, P., Ryu, C., Sharifi, V. & Swithenbank, J., Effect of process parameters on pelletisation of herbaceous crops. *Fuel*, **88**, pp. 1491–1497, 2009.
[87] Sultana, A., Kumar, A. & Harfield, D., Development of agri-pellet production cost and optimum size. *Bioresour Technol*, **101**, pp. 5609–5621, 2010.
[88] Verma, V.K., Bram, S., Delattin, F., Laha, P., Vandendael, I., Hubin, A. & de Ruyck, J., Agro-pellets for domestic heating boilers: Standard laboratory and real life performance. *Appl Energy*, **90**, pp. 17–23, 2012.
[89] Larsson, S.H., Thyrel, M., Geladi, P. & Lestander, T.A., High quality biofuel pellet production from pre-compacted low density raw materials. *Bioresour Technol*, **99**, pp. 7176–7182, 2008.

[90] Arshadi, M., Gref, R., Geladi, P., Dahlqvist, S.-A. & Lestander, T., The influence of raw material characteristics on the industrial pelletizing process and pellet quality. *Dimethyl Ether Spec Sect*, **89**, pp. 1442–1447, 2008.
[91] PelHeat, *Biomass Pellet Production Guide*. 2010.
[92] Zamorano, M., Popov, V., Rodríguez, M.L. & García-Maraver, A., A comparative study of quality properties of pelletized agricultural and forestry lopping residues. *Renew Energy*, **36**, pp. 3133–3140, 2011.
[93] Rhén, C., Gref, R., Sjöström, M. & Wästerlund, I., Effects of raw material moisture content, densification pressure and temperature on some properties of Norway spruce pellets. *Fuel Process Technol*, **87**, pp. 11–16, 2005.

CHAPTER 2

Factors Affecting Pellet Quality

Angela Garcia-Maraver[1] & Manuel Carpio[2]
[1] *Department of Civil Engineering, University of Granada, Campus de Fuente Nueva, 18071 Granada, Spain.*
[2] *Department of Building Construction, Campus de Fuente Nueva, 18071 Granada, Spain.*

Abstract

The quality of pellets depends on two general factors: (1) the characteristics of the raw material itself, directly related to its chemical composition and (2) the operative variables of the pelletization process, which influence the physical and mechanical properties of the pellets produced. A suitable combination of these two factors is necessary to obtain a qualified energy product. The European Union has established standards, guidelines and thresholds for solid biofuels in order that they can be properly used and marketed as renewable fuels.
Keywords: Composition, properties, quality, standards.

1 Introduction

The quality of pellets is determined by the end user's requirements in terms of the heating system and the handling properties [1]. In general, pellet quality depends on the properties of the feedstock – biomass type, moisture content and particle/fibre size – and management of the manufacturing process, including operating conditions, pelletizer type and binding agent [1–6].

These parameters involve high sensitivity when the feedstock changes. A better knowledge of the mechanical, physical and chemical properties of the biomass used is, therefore, necessary to determine its thermal behaviour [1,6,7]. These properties are described below.

2 Biomass properties

2.1 Chemical properties of the raw materials

There are great differences in the chemical properties of the various types of biomasses. The composition of the raw material varies from species to species, and within the same species it depends on the habitat, age and location in the tree (trunk, branches, top, leaves and roots) [8].

2.1.1 Chemical composition

Biomass has a number of constituents and complex structures. The varying amounts of organic and inorganic elements condition the final quality of the biomass as a fuel:

(a) Because of its organic nature, carbon, oxygen, nitrogen and hydrogen are the main components of biomass [9] and determine the calorific value of the raw material [10].
(b) Hemicellulose, cellulose, lignin and extractive matter are major constituents of biomass present in different ratios and structures in different biomass species. The mass loss during the devolatilization of biomass depends on these main components, which decompose at different temperature ranges [11].
(c) Sulphur and chlorine can reduce the melting point of the ash, causing operational problems in combustion, such as formation of deposits (slagging) in the furnace [12]. Their corrosive character can also reduce the protective oxide layer in the furnace.
(d) Potassium content influences the ash melting behaviour and aerosol formation. Increased potassium leads to a decreased ash melting point, which can cause slag and hard deposit formation in the furnace and boiler. Moreover, high potassium concentrations raise the amount of aerosols formed during combustion and thus cause fouling in boilers and fine particulate emissions [13].

2.1.2 Calorific value

The calorific value of biomass is an indication of the energy chemically bound in it. This is a critical property of a fuel because it determines the energy value. The design and control of a biomass combustor depend largely on the calorific value of the biomass fuel [14].

2.1.3 Moisture content

Water plays a key role in the pelletizing process [7,15,16]. It is the single most important factor influencing pellet quality. In addition to acting as a binding agent, affecting mechanical durability and the presence of fines, water is a lubricant that lowers the friction in the die, resulting in low bulk density and low energy consumption [7].

2.1.4 Ash content

One problem associated with biomass combustion resides in the ash content of the material. The quantity and quality of ash produced during the biomass combustion process are strongly influenced by the properties of the biomass and the combustion technology, including the characteristics of the furnace, the temperature of the process and the ash extraction system. Inorganic species in biomass fuels, such as alkali oxides and salts, can intensify agglomeration, deposition and corrosion problems on the heat transfer surfaces of boilers [17]. Despite the obvious benefits of using biomass as a fuel source, technical difficulties related to the inorganic part of biomass fuels can interfere with the accessibility of biomass combustion systems. Higher ash content means lower heating value of the pellets and a greater risk of sintering. Since this also negatively affects milling and pelleting equipment [1], the calculation and subsequent reduction of the ash content in biofuels are crucial to enhance their quality. It goes without saying that this is also an obstacle for the residential pellet market [3,18].

2.1.5 Volatile content

Compared with coal, biomass has a high volatile content. Though held to be an outstanding renewable energy source, biomass generates a number of pollutants, including particulate matter, unburned heavy hydrocarbons and nitrogen oxides [19]. These emissions increase in the case of biomass with high ash and nitrogen contents [20], but the type of burner, boiler or stove in use is also significant [21].

Like coal volatiles, biomass volatiles result from pyrolysis and are typically made up of H_2, H_2O, CO, CO_2, hydrocarbon gases, hydrocarbon liquids and polyaromatic hrdrocarbons (PAHs). If biomass is to be more widely used as an energy source, a greater understanding of this aspect is necessary. According to Williams *et al.* [22], biomass combustion modelling must account for prediction for gas phase flow and volatile combustion, including a detailed chemical mechanism for calculating pollutants of PAHs and soot particles. It is not yet known how biomass conversion mechanisms work at high temperatures, or specifically how the nature of the biomass affects the formation of PAHs and soot during combustion.

2.1.6 Thermal behaviour

The thermal decomposition of lignocellulosic materials involves a number of chemical reactions as well as physical stages such as heat and mass transfer [23].

Wood is a complex material, mainly composed of cellulose (40–60 wt%), hemicellulose (20–40 wt%) and lignin (10–25 wt%), in addition to extractives (terpens, tannins, fatty acids, resins, etc.), water and mineral matter [24].

Differences in the main chemical components of biomass (i.e. cellulose, hemicelluloses and lignin) bear a direct impact on their chemical reactivity. Given the heterogeneous nature of biomass, it is not easy to make generalizations about thermochemical treatment systems and their equipment [25–27].

When thermal decomposition takes place in an air atmosphere, the presence of oxygen causes additional complexities. It enhances the degradation of the material at low temperatures [28] and causes the appearance of gas-phase reactions between the volatile compounds released and oxygen; when the temperature is high enough, oxygen is moreover likely to promote the combustion of the char residue generated in the early stages of solid degradation [28]. This means that combustion and pyrolysis may coexist [29]. The kinetics of this thermal decomposition is more closely related to the combustion process.

The thermogravimetric analysis curve in combination with its corresponding derivative DTG peaks can indicate the number of stages of thermal degradation, which are directly related to the decomposition of the main components [29–31]. DTG is used to determine the characteristic temperatures such as the temperature of maximum mass loss rate, the initial decomposition temperature and the final decomposition temperature. Knowledge of the combustion characteristics is the basis for a better understanding of the thermal–chemical conversion of biomass.

2.2 Mechanical and physical properties of pellets

Because the mechanical and physical properties are related to the manufacturing process, the changes in the operational conditions of the pelletization affect these final parameters.

2.2.1 Durability

Mechanical durability (or abrasive resistance) is a quality parameter that can be defined as the ability of densified biofuels to remain intact when handled [32]. It is measured by the resistance of densified fuels towards shock or/and friction. Since the pellet is susceptible to mechanical wear, which leads to the production of fine particles or dust during transport and storage [4], the resistance test will indicate the dusting potential of the pellets during handling, transportation and storage. Dust emissions are not only an inconvenience for the consumer, they are also a health hazard [33]. Additionally, fine particles and dust can disturb the feeding systems of boilers and may lead to inhomogeneous combustion processes. Finally, dust may contribute to fire and explosion risks during handling, storage and transhipment [34].

2.2.2 Hardness

Hardness (or compressive or crushing resistance) designates the maximum crushing load a pellet can withstand before cracking or breaking [7]. The tensile strength is related to the adhesion forces between particles at all contact points in the agglomerate [6,35]. A compressive resistance test simulates the compressive stress due to the weight of the top pellets on the lower pellets during storage in bins or silos, and the crushing of pellets in a screw conveyor. This provides a quick measure of the quality of pellets as soon as they are produced by the pellet mill, and it helps to make adjustments in the pelleting process to improve the product quality.

2.2.3 Bulk density

The bulk density of pellets is a parameter used for piles of wood fuels, as voids among the woody pieces may be bigger or smaller depending on the size and shape of the pellets [36].

The bulky characteristic of non-densified biomass makes it difficult to transport over long distances and demands space for storage. The densification of biomass by pelleting can significantly increase the bulk density of biomass from 40–250 kg/m^3 up to 600–800 kg/m^3 [7,15] and thus improve storability and reduce transportation costs [7,37]. Moreover, bulk density is an important consideration since the fuel is fed by volume, not weight, meaning that variations in bulk density can cause considerable variations in combustion efficiency.

2.2.4 Particle density

Particle density is defined as the ratio of the mass and the volume of a sample including pore volume. Its value is affected by the particle size, compression strength, and protein and moisture content [38].

This property has an influence on the bulk density and on the combustion properties of pellets (heat conductivity, burning time and rate of degasification) [39].

2.2.5 Size of the pellets

Wood and wood-based materials interact with a humid environment by adsorbing and desorbing moisture. One factor that may limit the hygroscopicity is the hard circumference surface around the pellet. This surface is a result of manufacturing under the high pressures needed to produce the pellet. However, the ends of the pellets are exposed to the humid environment and any moisture uptake by the pellet would be from the ends [34,40], so an important parameter is the length/diameter ratio.

In this context, the length/diameter ratio is also of great relevance when pneumatic feeding systems are used, due to the fact that even a single long pellet could block the transport pipe of such systems [13].

3 European pellet quality standards

Apart from the energy content of pellets, end consumers are mainly concerned with two problem areas: (1) emissions generated by pellet combustion and (2) the use of pellets in boilers and stoves. This is understandable since emissions and low-quality biofuels can damage combustion equipment and produce undesired effects such as slagging and corrosion [1]. Furthermore, the content of fine particles in the fuel – which is extremely dependent on mechanical durability as well as logistics and storage – can disturb the regulation of highly automated heating systems or interrupt automated fuel feeding [2]. Fine particles burn more quickly, and these higher temperatures can favour ash melting.

Many European countries have consequently developed standards for the quality, storage, transport and combustion of densified biomass fuels. The parameters and guidelines address: particle and bulk density, moisture content, crushing resistance or hardness, particle number, particle size (length and diameter), chemical composition, ash content and heating value. As trade between countries becomes more widespread, there is a need for international standards to facilitate the purchase and sale of biomass fuels. Accordingly, the European Committee for Standardization (CEN) received a mandate from the European Commission to develop standards for solid biofuels within the Technical Committee (TC) 335 Solid Biofuels.

Still, current national standards regarding pellet quality are very heterogeneous. Because of the differences in quality control parameters and guidelines, we drew a comparison of the parameters and the limit values for densified biomass fuels, as defined in different European standards.

3.1 Pellet quality: standards and certification

3.1.1 Austrian standard: ÖNORM M 7135

The Austrian standard for energy pellets is the ÖNORM M 7135, which establishes requirements and test specifications for compressed wood or compressed bark in its natural state – pellets and briquettes [41]. It, furthermore, includes requirements for non-wood biomass pellets (i.e. M7139 for energy grains, C 4000 for Miscanthus pellets and C 4002 for straw pellets). The main parameters and guidelines of the ÖNORM M 7135 are summarized in Table 1. Compliance with this standard is certified by the ÖNORM tested label for high-quality pellets. Other pellet-related standards in Austria are those pertaining to pellet logistics (M 7136) and pellet storage tanks (M 7137).

3.1.2 Swedish standards: SS 187120

Sweden was one of the first countries to establish pellet quality standards in its SS 187120 [42]. This standard includes three categories of pellet quality, depending on the parameter limits considered. The physical, mechanical and chemical parameters, as well as the guidelines under this standard, are listed in Table 2.

3.1.3 German standards: DIN 51731 and DIN EN 15270

Germany has two standards: DIN 51731 [43] for pellets and briquettes, and DIN EN 15270 [44] for high-quality pellets. The first standard for compressed wood

Table 1: Parameters and guidelines under Austrian standards for energetic pellets [41].

Parameter		Limits
	Diameter (mm)	4–10
Physical	Length (mm)	<5D
	Particle density (kg/dm^3)	<1.12
Mechanical	Durability* (%)	<2.3
	Moisture content (%)	<10
	Ash content (%)	<0.5
	Heating value (kcal/kg)	>4302
Chemical	N (%)	<0.3
	S (%)	<0.04
	Cl (%)	<0.02
	Additives (%)	<2

*Durability is defined in terms of percentage of fines after testing.

Table 2: Parameters and guidelines included in Swedish standards for biofuels and peat [42].

Parameter		Pellet category		
		Group 1	Group 2	Group 3
Physical	Diameter (mm)	<25	<25	<25
	Length (mm)	<4D	<5D	<5D
	Bulk density (kg/dm^3)	>600	>500	>500
	Moisture content (%)	<10	<10	<10
Mechanical	Durability (%)*	<0.8	<1.5	<1.5
Chemical	Ash content (%)	<0.7	<1.5	<1.5
	Heating value (kcal/kg)	>4039	>4039	>3609
	S (%)	<0.08	<0.08	–
	Cl (%)	<0.03	<0.03	–

*Durability is defined in terms of percentage of fines after testing.

was introduced in Germany in 1996. The use of this standard in the small-scale pellet heating sector is limited because of its high maximum ash content (<1.5%) and the absence of an abrasion limit (Table 3).

The DINplus certification, based on the standard for high-quality wood pellets, was subsequently developed. This standard has contributed significantly to the success of the residential pellet market in Germany. In fact, today it is the most important quality certification for wood pellets worldwide. Its parameters and limit values are listed in Table 3.

3.1.4 Italian standard: CTI-R04/05

The Italian standard CTI-R04/05 establishes the quality parameters for energy pellets from biomass [45]. It classifies pellets in the following three categories, depending on their composition:

A.1. Deciduous and conifer tree trunk without bark; untreated wood from wood-yielding industry; untreated wood and wood without bark after being used; a mixture of these materials.
A.2. Raw materials not included in category A.1; untreated herbaceous biomass; a mixture of these materials.
A.3. Raw materials not included in category A.2.

Table 4 summarizes the parameters, values and guidelines for each pellet category.

3.1.5 French recommendation: ITEBE

Even though France has no official standard for wood pellets, the government has established certain quality controls [46], which classify pellets according to their usage context (stove, boiler, large boiler and incinerator). Table 5 summarizes the control and parameter values of these recommendations.

Table 3: Parameters and guidelines included in German standards [43,44].

Parameter		DIN 51731	DIN EN 15270
Physical	Diameter (mm)	4–10	–
	Length (mm)	<5D	<5D
	Particle density (kg/dm^3)	<1.2	<1.2
Mechanical	Durability (%)*	–	<2.3
Chemical	Moisture content (%)	<12	<10
	Ash content (%)	<1.5	<0.5
	Heating value (kcal/kg)	3705–4661	>4302
	N (%)	<0.3	<0.3
	S (%)	<0.08	<0.04
	Cl (%)	<0.03	<0.02
	Additives (%)	–	<2
	As (mg/kg)	<0.8	<0.8
	Cd (mg/kg)	<0.5	<0.5
	Cr (mg/kg)	<8	<8
	Cu (mg/kg)	<5	<5
	Hg (mg/kg)	<0.05	<0.05
	Pb (mg/kg)	<10	<10
	Zn (mg/kg)	<100	<100

*Durability is defined in terms of percentage of fines after testing.

Table 4: Parameters and guidelines included in Italian recommendation [45].

Parameter		Pellet categories		
		A.1	A.2	A.3
Physical	Diameter (mm)	6 ± 0.5 to 8 ± 0.5	6 ± 0.5 to 8 ± 0.5	10 ± 0.5 to 25 ± 1.0
	Length (mm)	–	<5D	–
	Bulk density (kg/dm^3)	620–720	600–720	≥550
Mechanical	Dust emission (%)	≤1	≤1	–
Chemical	Moisture content (%)	≤10	<10	≤15
	Ash content (%)	≤0.7	<1.5	–
	Heating value (kcal/kg)	>4039	>3870	–
	N (%)	≤0.3	≤0.3	–
	S (%)	≤0.5	≤0.5	–
	Cl (%)	≤0.03	–	–
	Additives (%)	banned	–	–

Table 5: Parameters and guidelines of French recommendation [46].

Parameter		Type of combustion installation			
		Stove	Boiler	Big boiler	Incinerator
Physical	Diameter (mm)	6 ± 1	8–10 ± 1	>16	>16
	Length (mm)	10–30	10–15	>16	>16
	Bulk density (kg/dm^3)	>650	>650	>580	>580
	Particle density (kg/dm^3)	1.2–1.4	1.2–1.4	–	–
Chemical	Moisture content (%)	<10	<10	–	–
	Ash content (%)	<10	<10	–	–
	Heating value (kcal/kg)	>4052	>4052	–	–
	N (%)	<0.3	<0.3	–	–
	S (%)	<0.08	<0.08	–	–
	Cl (%)	<0.3	–	–	–
	Na (ppm)	<300	–	–	–

3.1.6 European standard committee CEN/TC 335

Finally, the European Standard Committee CEN/TC 335 is the technical committee that has developed the draft standard to describe all forms of solid biofuels in Europe, including wood chips, wood pellets and briquettes, logs, sawdust and straw bales.

The result is the experimental technical specification for solid biofuels [32], which includes and specifies the relevant terminology, definitions and descriptions.

The guidelines and pellet classification of the European standard [47] are different from those of national standards. It specifically classifies pellets in three general categories, depending on the origin of the raw materials: (1) woody biomass; (2) herbaceous biomass; (3) biomass from fruit or a mixture of the previous types of biomass.

Each category includes various subcategories, specified to further levels. For example, the category of woody biomass includes the following four subcategories: (1.1) forest, plantation and other virgin wood; (1.2) by-products and residues from wood processing industry; (1.3) used wood; (1.4) blends and mixtures. Subcategory 1.1 has seven subcategories: (1.1.1) whole trees without roots; (1.1.2) stemwood; (1.1.3) logging residues; (1.1.4) stumps/roots; (1.1.5) bark (from forestry operations; (1.1.6) segregated wood from gardens, parks, roadside maintenance, vineyards and fruit orchards; (1.1.7) blends and mixtures. Finally, each of these subcategories also contains various subdivisions. For instance, category 1.1.1 is subdivided as follows: (1.1.1.1) deciduous; (1.1.1.2) coniferous; (1.1.1.3) bushes; (1.1.1.4) blends and mixtures. Each parameter likewise has various subdivisions, depending on their limit values (Table 6). Labels should provide informa-

Table 6: Classification of parameters included in the European Standard including specifications for solid biofuels [47].

Parameter	Classification
Size (diameter and length) (mm)	D06: D ≤ 6 ± 0.5 and L ≤ 5D
	D08: D ≤ 8 ± 0.5 and L ≤ 4D
	D10: D ≤ 10 ± 0.5 and L ≤ 4D
	D12: D ≤ 12 ± 1.0 and L ≤ 4D
	D25: D ≤ 25 ± 1.0 and L ≤ 4D
Moisture content (%)	M10: ≤10%
	M15: ≤15%
	M20: ≤20%
Ash content (%)	A0.7: ≤0.7%
	A1.5: ≤1.5%
	A3.0: ≤3%
	A6.0: ≤6%
	A6.0+: >6%
N (%)	N0.3: ≤0.3%
	N0.5: ≤0.5%
	N1.0: ≤1%
	N3.0: ≤3%
	N3.0+: >3%
S (%)	N0.05: ≤0.05%
	N0.08: ≤0.08%
	N0.1: ≤0.1%
	N0.2+: >0.2%
Cl (%)	CL0.03: ≤0.03
	CL0.07: ≤0.07
	CL0.1: ≤0.1
	CL0.1+: >0.1
Durability*	DU97.5: ≥97.5
	DU95.0: ≥95
	DU90: ≥90
Fines content (% < 3.15 mm)	F1.0: ≤1%
	F2.0: ≤2%
	F2.0+: >2%
Bulk density (kg/m^3)	Recommended value should be included by manufacturer
Heating value (kcal/kg)	Recommended value should be included by manufacturer
Additives	Binding materials and ash inhibitory should be included in the label

*Durability is defined in terms of percentage of whole pellets after testing.

tion about the characteristics of each type of pellet. Furthermore, the annexes of this standard include recommendations for the potential uses of pellets. For example, the recommended parameter values for pellets used in domestic heating are stricter because (1) domestic equipment does not employ advanced cleaning and control mechanisms; (2) the equipment is not handled by professional staff; (3) the combustion of pellets takes place in inhabited areas.

3.2 Comparative analysis of European pellet quality standards

The standards for pellet quality developed by various European countries differ not only in their basic requirements but also in their limit values and their mode of application to the pellets. Pellet size, moisture content, heating value and ash content are the pellet quality parameters that appear most frequently in standards and recommendations. However, pellet size (diameter, length and/or length/diameter ratio) is the only parameter that is included in all the norms.

Most of the standards specify moisture content limit values except in the case of large boilers and incinerators under the French recommendations [46]. Bulk density is also included in practically all the norms, with the exception of the German standards [43,44].

Nitrogen and sulphur limit values are provided in the Austrian norm [41], the German standards [43,44], the Italian standard (for A1 and A2 pellets) [45] and the French recommendations (for the use of pellets in stove and heating) [46].

Heating limit values appear in all the norms except in the Italian standard for pellet type A3 [45].

Most standards also include ash content limit values, except in the case of A3 pellets in the Italian standard [45] and pellets for large boilers and incinerators under the French recommendations [46].

Finally, there are other parameters that appear in only a few of the standards. For example, the German standards include limit values for the As, Cd, Cr, Cu, Hg, Pb and Zn content of pellets. This is true even in standard DIN 51731 for lower quality pellets [43,44].

Particle density is specified in the German standards and the French recommendations (i.e. for the use of pellets in stove and boilers) [43,44,46]. The German standard DinEN 15270 [44] is the only one including limit values for additives, though this parameter is restricted to higher quality pellets.

In reference to the limit values of the various parameters, no important differences were observed, except in the case of ash content, durability and heating value. In these cases, the German standard DinEN 15270 [44] and the Austrian norm [41] were found to be the most restrictive.

Finally, there are norms that establish quality parameters according to a pre-established classification of pellets. For example, the Swedish standard specifies three categories of pellet quality depending on the limit values set. The set of values for group 1 defines pellets of the highest quality; the set for group 2 defines pellets of medium quality; and the set for group 3, pellets of lesser quality [42]. The classification of pellets in the Italian recommendation [45] is based on the origin of the raw material, and the limit values are more restrictive for pellets made with higher quality raw materials, such as deciduous and conifer tree trunks without bark, which belong to group A.1. For pellets made of untreated herbaceous biomass (group A.2), a lower heating value is specified, and the ash content may be higher. Group A.3 has even fewer restrictions, and the guidelines only specify values for the diameter, bulk density and moisture content of the pellets. French quality controls establish four groups of pellets depending on their final use. Accordingly, pellets of higher quality should be used in stoves or boilers, while less exacting requirements are applicable to pellets used in large boilers and incinerators [46].

In light of the heterogeneous nature of national pellet quality standards, the experimental technical specification EN 14961-1 [47] for solid biofuels created a set of guidelines for pellet quality and use. It also elaborated a very specific categorization of pellets, based on the classifications under national norms, and classifying pellets depending on the origin of the raw materials.

4 Conclusions

In the vast majority of the European countries, there are very few regulations for the manufacture and quality of pellets. All too often, pellets merely come under the jurisdiction of very general biomass laws. The only European countries with official standards for compacted biomass fuels are Austria, Sweden and Germany. A few other countries, such as France and Italy, have published recommendations or guidelines.

The comparative analysis of these standards and recommendations for pellets highlights significant differences in their quality parameters as well as in their guidelines. National standards that specify pellet quality requirements at an individual level have not been accepted by all stakeholders in Europe, and this has contributed to the development of national pellet types, whose characteristics depend on the country where they are manufactured. The situation will no doubt change when the application of the European standard for solid biofuels becomes official. A common standardization will be beneficial for Europe. By eliminating an important barrier to the general development of the European market for pellet and combustion technology, it will increase the international pellet trade in particular.

References

[1] Lehtikangas, P., Quality properties of pelletised sawdust, logging residues and bark. *Biomass and Bioenergy*, **20**, pp. 351–360, 2001.
[2] Li, Y. & Liu, H., High-pressure densification of wood residues to form an upgraded fuel. *Biomass and Bioenergy*, **19**, pp. 177–186, 2000.
[3] Rhén, C., Gref, R., Sjöström, M. & Wästerlund, I., Effects of raw material moisture content, densification pressure and temperature on some properties of Norway spruce pellets. *Fuel Process Technol*, **87**, pp. 11–16, 2005.

[4] Temmerman, M., Rabier, F., Jensen, P.D., Hartmann, H. & Böhm, T., Comparative study of durability test methods for pellets and briquettes. *Biomass and Bioenergy*, **30**, pp. 964–972, 2006.
[5] Arshadi, M., Gref, R., Geladi, P., Dahlqvist, S.-A. & Lestander, T., The influence of raw material characteristics on the industrial pelletizing process and pellet quality. *Dimethyl Ether Spec Sect*, **89**, 1442–1447, 2008.
[6] Gilbert, P., Ryu, C., Sharifi, V. & Swithenbank, J., Effect of process parameters on pelletisation of herbaceous crops. *Fuel*, **88**, pp. 1491–1497, 2009.
[7] Kaliyan, N. & Vance Morey, R., Factors affecting strength and durability of densified biomass products. *Biomass and Bioenergy*, **33**, pp. 337–359, 2009.
[8] Orfão, J.J.M., Antunes, F.J.A. & Figueiredo, J.L., Pyrolysis kinetics of lignocellulosic materials – three independent reactions model. *Fuel*, **78**, pp. 349–358, 1999.
[9] Byrne, C.E. & Nagle, D.C., Carbonization of wood for advanced materials applications. *Carbon N Y*, **35**, pp. 259–266, 1997.
[10] Telmo, C., Lousada, J. & Moreira, N., Proximate analysis, backwards stepwise regression between gross calorific value, ultimate and chemical analysis of wood. *Bioresour Technol*, **101**, pp. 3808–3815, 2010.
[11] Prins, M.J., Ptasinski, K.J. & Janssen, F.J.J.G., Torrefaction of wood: part 1. Weight loss kinetics. *J Anal Appl Pyrolysis*, **77**, pp. 28–34, 2006.
[12] Öhman, M. & Nordin, A., A new method for quantification of fluidized bed agglomeration tendencies: a sensitivity analysis. *Energy Fuels*, **12**, pp. 90–94, 1998.
[13] Obernberger, I. & Thek, G., Physical characterisation and chemical composition of densified biomass fuels with regard to their combustion behaviour. *Biomass and Bioenergy*, **27**, pp. 653–669, 2004.
[14] Erol, M., Haykiri-Acma, H. & Küçükbayrak, S., Calorific value estimation of biomass from their proximate analyses data. *Renew Energy*, **35**, pp. 170–173, 2010.
[15] Mani, S., Tabil, L.G. & Sokhansanj, S., An overview of compaction of biomass grinds. *Powder Handl Process*, **15**, pp. 160–168, 2003.
[16] Nielsen, N.P.K.M., Gardner, D.J., Poulsen, T. & Felby, C., Importance of temperature, moisture content, and species for the conversion process of wood residues into fuel pellets. *Wood Fiber Sci*, **41**, pp. 414–425, 2009.
[17] Tortosa Masiá, A.A., Buhre, B.J.P., Gupta, R.P. & Wall, T.F., Characterising ash of biomass and waste. *Impacts Fuel Qual Power Prod*, **88**, pp. 1071–1081, 2007.
[18] Öhman, M., Boman, C., Hedman, H., Nordin, A. & Boström, D., Slagging tendencies of wood pellet ash during combustion in residential pellet burners. *Biomass and Bioenergy*, **27**, pp. 585–596, 2004.
[19] Koziński, J.A. & Saade, R., Effect of biomass burning on the formation of soot particles and heavy hydrocarbons. An experimental study. *Fuel*, **77**, pp. 225–237, 1998.

[20] Johansson, L., Emission characteristics of modern and old-type residential boilers fired with wood logs and wood pellets. *Atmos Environ*, **38**, pp. 4183–4195, 2004.

[21] Olsson, M., Kjällstrand, J. & Petersson, G., Specific chimney emissions and biofuel characteristics of softwood pellets for residential heating in Sweden. *Biomass and Bioenergy*, **24**, pp. 51–57, 2003.

[22] Williams, A., Jones, J.M., Ma, L. & Pourkashanian, M., Pollutants from the combustion of solid biomass fuels. *Prog Energy Combust Sci*, **38**, pp. 113–137, 2012.

[23] Bilbao, R., Mastral, J.F., Aldea, M.E. & Ceamanos, J., Kinetic study for the thermal decomposition of cellulose and pine sawdust in an air atmosphere. *J Anal Appl Pyrolysis*, **39**, pp. 53–64, 1997.

[24] McKendry, P., Energy production from biomass (part 1): overview of biomass. *Rev Issue*, **83**, pp. 37–46, 2002.

[25] Barton, G.M., *Organic Chemistry from Biomass*, CRC Press: Boca Raton, Florida, pp. 247–275, 1981.

[26] Pereira, H. & Sardinha, R., Chemical composition of *E. globulus* (Labill). *Appita*, **37**, pp. 661–664, 1984.

[27] Pereira, H., Variability in the chemical composition of plantation eucalypts (*Eucalyptus globulus Labill*). *Wood Fiber Sci*, **20**, pp. 82–90, 1988.

[28] Shafizadeh, F. & Bradbury, A.G.W., Thermal degradation of cellulose in air and nitrogen at low temperatures. *J Appl Polym Sci*, **23**, pp. 1431–1442, 1979.

[29] Liu, N.A., Fan, W., Dobashi, R. & Huang, L., Kinetic modeling of thermal decomposition of natural cellulosic materials in air atmosphere. *J Anal Appl Pyrolysis*, **63**, pp. 303–325, 2002.

[30] Grioui, N., Halouani, K., Zoulalian, A. & Halouani, F., Thermogravimetric analysis and kinetics modeling of isothermal carbonization of olive wood in inert atmosphere. *Thermochim Acta*, **440**, pp. 23–30, 2006.

[31] Marcilla, A., Gómez-Siurana, A., Gomis, C., Chápuli, E., Catalá, M.C. & Valdés, F.J., Characterization of microalgal species through TGA/FTIR analysis: application to nannochloropsis sp. *Thermochim Acta*, **484**, pp. 41–47, 2009.

[32] EN14588. Solid biofuels. Terminology, definitions and descriptions. 2011.

[33] Vinterbäck, J., Pellets 2002: the first world conference on pellets. *Biomass and Bioenergy*, **27**, pp. 513–520, 2004.

[34] Lehtikangas, P., Storage effects on pelletised sawdust, logging residues and bark. *Biomass and Bioenergy*, **19**, pp. 287–293, 2000.

[35] Pietsch, W., *Agglomeration Processes – Phenomena, Technologies, Equipment*, Wiley-VCH: Weinheim, 2002.

[36] EN15103. Solid biofuels. Determination of bulk density. 2009.

[37] Holm, J.K., Henriksen, U.B., Hustad, J.E. & Sorensen, L.H., Toward an understanding of controlling parameters in softwood and hardwood pellets production. *Energy Fuels*, **20**, pp. 2686–2694, 2006.

[38] Mani, S., Tabil, L.G. & Sokhansanj, S., Effects of compressive force, particle size and moisture content on mechanical properties of biomass pellets from grasses. *Biomass and Bioenergy*, **30**, pp. 648–654, 2006.
[39] Böhm, T. & Hartmann, H., Measuring particle density of wood pellets. *Proc. Second world Conf. Technol. Exhib. biomass energy, Ind. Clim. Prot.*, Rome, Italy, 2004.
[40] Hartley, I.D. & Wood, L.J., Hygroscopic properties of densified softwood pellets. *Biomass and Bioenergy*, **32**, pp. 90–93, 2008.
[41] ÖNORM M7135. Compressed wood or compressed bark in natural state, pellets and briquettes. Requirements and test specifications. 2003.
[42] SS187120. Biofuels and peat, fuel pellets. Classification (Swedish Standards Institution). 1998.
[43] DIN51731. Testing of solid fuels, compressed untreated wood. Requirements and testing (Deutsches Institut für Normung). 1996.
[44] DinEN15270. Pellet burners for small heating boilers. Definitions, requirements, testing, marking (Deutsches Institut für Normung). 2007.
[45] CTI-R04/5. Recommendation: solid biofuels. Pellet characterization for energetic purposes. 2004.
[46] ITEBE. France does not have an official standard but a set of quality controls has been developed. 2009.
[47] EN14961-1. Solid biofuels. Fuel specification and classes. Part 1: general requirements. 2010.

CHAPTER 3

Physical Transformation Processes of Forest Biomass: Chipping, Bundling, Drying and Control of Particle Size Distribution

Daniel J. Vega-Nieva & Luis Ortiz
Forestry School, University of Vigo, Campus A Xunqueira, 36005 Pontevedra, Spain.

Abstract

Forest biomass is characterized by physical properties such as high moisture content, low density and heterogeneous particle size, which condition its direct use for combustion. Hence, processes of physical transformation are previously required to obtain products of higher energetic value, reducing transport costs and allowing for homogeneous boiler feeding. The biomass harvesting processes, as well as the logistic options for chipping and bundling, are described in this chapter, which offers a summary of the costs of such operations measured in forest biomass logistic studies in Spain. Studies on natural versus controlled drying of chips and bundles are also discussed, and general recommendations for the management of moisture content and the prevention of self-ignition risk during storage are given. Finally, experiments to date involving the screening and size reduction of forest biomass are summarized.
Keywords: Forest biomass, chipping, bundling, drying, screening, size reduction.

1 Forest Biomass Harvesting and Transportation

There are two main types of forest biomass sources: residual forestry biomass and energy crops. The difference resides in the harvesting and transportation costs due to their logistic chains, and they will be analysed separately.

1.1 Harvesting and transportation costs of forest residual biomass

Several experiments in forest residue collection for bioenergy compare the productivity and costs of field chipping, forest track chipping and bundling strategies on different coniferous and deciduous forest harvesting [1,2].

- Residue chipping in the forest track involves accumulating residues with a forwarder, and allowing the moisture content of residues to decrease by natural drying at the forest track piles. According to the Spanish IDAE [3] guidelines, this is the most economical system for large forests. It is, moreover, the option favoured by European Nordic countries with experience in biomass residue collection [4].
- The bundling system, on the other hand, is limited by a high machinery cost of 98 €/h [5] and the need for posterior chipping of residues; its use is on the decline with respect to chipping systems.

A series of experiments in collection and treatment of forest residues (*Pinus pinaster* and *Eucalyptus globulus*) were conducted in Galicia, NW Spain, by Sanz and Piñeiro [1] to compare the productivity of diverse collection systems. Three different residue situations were considered (Fig. 1):

- *Situation 1*: Residues scattered on the field.
- *Situation 2*: Residues grouped in small piles on the field.
- *Situation 3*: Residues grouped in large piles at the forest track.

The piling of residues (situations 2 and 3) reduced biomass moisture content and generally decreased collection costs as compared to situation 1, where residues were not grouped previous to collection.

The operation productivity, residue moisture content (MC) and residue collection costs (in 2003) obtained in this study are summarized in Table 1 for the four different scenarios (forwarding distance = 300 m) considered in the study (Fig. 2):

- SC. 1.a: Chipharvester Bruks 803 CT on Valmet 892 forwarder.
- SC. 2.a: Pezzolato PHT 1200 chipper on Timberjack 1210 B forwarder.
- SC. 3.a: Fiberpac 370 B Bundler on Timberjack 1210 B forwarder.
- SC. 4 a: Balapress Bundler.

Figure 1: Types of biomass collection logistics studied by Sanz and Piñeiro [1] in NW Spain. From left to right: situation 1: residues scattered on the field; situation 2: residues grouped in small piles on the field; situation 3: residues grouped in large piles at the forest track.

Figure 2: Mobile chippers and bundlers utilized in the Sanz and Piñeiro [1] study in NW Spain.

Table 1: Summary of Sanz and Piñeiro [1] biomass collection experiences in Galicia, NW Spain.

Scenario	Species	MC (%)	Operation productivity (green tonnes/h)	Residue collection costs (€/green tonne)
SC.1a	*Pinus* situation 2	35	10.7	8.9
	Pinus situation 3	31	14	6.8
	Eucalyptus situation 1	47	9	10.6
	Eucalyptus situation 3	26	12.6	7.6
SC.2a	*Pinus* situation 3	31	13.6	7.6
SC.3a	*Pinus* situation 2	35	9.6	8.3
	Pinus situation 3	31	9.3	8.6
	Eucalyptus situation 1	47	9.9	8
	Eucalyptus situation 3	26	13.1	6.1
SC. 4a	*Pinus* situation 3 (average production)	31	3.4	28.8
	Pinus situation 3 (maximum production)	31	4.9	20

These results were updated to prices of 2007 by Pedras [2] to include an estimation of transport cost while assuming an optimal organization of biomass collection and treatment and a transport distance of 100 km. In this case, the cost for the biomass plant would range between 23 and 28 €/green tonne for the field chipping with a chip harvester, and between 27 and 32.8 €/green tonne for the Fiberpac collection and plant chipping (Table 2). These costs include neither industrial profit nor indirect costs.

The Polytechnic University of Madrid, in cooperation with CESEFOR, conducted several studies of biomass residue collection in different types of harvests of various *Pinus*, *Quercus* and *Populus* stands [4,6]. Table 3 summarizes the total costs obtained in the cases studied (SC. 1b: whole tree harvesting and chipping at forest; SC. 2b: residue collection and chipping at forest; SC. 3b: mobile chipping at forest; SC. 4b: bundling and chipping at plant), including industrial benefit and indirect costs.

1.2 Harvesting and transportation costs of woody energy crops

Woody energy crops are generally established at higher densities than conventional forestry plantations, and can be managed in short rotation coppice systems, with rotations of 2–3 years generally for willow (*Salix* sp.) and 2–5 years for poplar [7]. Poplar energy crops were established commercially in Italy, where 5,000 hectares are managed for chip production, and commercial willow energy crops have been established in Sweden and the UK, based on specifically bred clones for biomass production [8].

Table 2: Total costs of the highest productivity scenarios from the CIS-Madera experiences in Galicia [2].

Scenario	Costs of biomass at plant (€/green tonne)			
	Residue collection	Transport (100 km)	Chipping at plant	Total cost
SC 1.a.	11.9–14.5	11–13.5	–	23–28
SC 3.a.	13.5–16.5	10–12	3.5–4.5	27–33

Table 3: Summary of costs obtained in forest biomass residue utilization experiences in Castilla-Leon, Spain.

Scenario	Species	MC (%)	Costs (€/green tonne)	Distance to plant (km)
SC. 1b	*Quercus pyrenaica*	42	54	100
	Pinus silvestris	56.2	34.2	40
SC. 2b	*Populus*	45.5	40.9	100
SC. 3b	*Pinus pinaster*	35	27.4	40
SC.4b	*Populus*	50	38	50
	Pinus	50	33–47	140–150

Table 4: Total biomass costs at plant for Populus energy crop [16].

Yield, DM tonne/ha/year	System*	Costs (€/green tonne)			
		Production and harvesting	Transport	Chipping	Total costs
20	W. tree	4.8	2.93	2.26	9.9
	Chips	5.2	4.24		9.4
15	W. tree	6.4	2.93	2.26	11.5
	Chips	7.0	4.24		11.2
13.5	W. tree	7.3	2.93	2.26	12.4
	Chips	7.7	4.24		11.9
9	W. tree	10.9	2.93	2.26	16.0
	Chips	11.6	4.24		15.8

*W. tree: whole tree harvesting (MC = 50%, transport at 20%); chips: tree chipping (MC = 50%).

Information on current research on varieties selection and management optimization of woody energy crops can be found in the literature [7,9–11].

In Spain, the studies of Ciria *et al.* [12–14] and San Miguel and Montota [15] have determined the yield potential of high-density poplar energy crop plantations for biomass productions, with yields of 10–18 dry matter tonne/ha in rotations of 3–6 years at densities of 5,000–10,000 stems/ha and higher. A recent study [16] analysed the economic feasibility of poplar energy crop plantations for bioenergy production, considering a life cycle of 16 years in three harvests at 5-year coppice intervals, comparing whole tree harvesting and field chipping systems.

2 Forest Biomass Drying

Woody biomass commonly has high moisture content after harvest, which implies limitations on its energetic utilization that includes:

- Increase of harvest and transport costs.
- Decrease of the efficiency in the physical transformation stages.
- Low efficiencies in the thermochemical conversion processes.
- Increase of CO_2 and NO_x emissions during combustion.
- Unstable combustion.
- Limitations or avoidance of its use as a fuel.

To overcome these limitations, moisture content in forest biomass must be decreased up to values of 20%–30% by natural or artificial drying.

Figure 3: Open air natural drying of wood chips at the CIEMAT Renewable Energy Centre in Soria, central Spain [17,18].

2.1 Natural drying

Natural drying is performed when the climatic conditions allow for the loss of moisture controlled by the environmental temperature and relative humidity conditions, under the absence of precipitation. In the case of forest biomass residues from harvest, there are two possibilities for natural drying: to conduct the drying on the field directly after harvest or to conduct the drying after chipping. In cold climate countries, for instance in Northern Europe, harvest residues are often piled in the field until attaining a field stockage of biomass that will dry under environmental conditions in the field (Fig. 3).

2.1.1 Thermogenesis process

During the storage of woody biomass, thermogenesis may occur due to the living cells of the parenchyma, the biological activity of microbial bacteria and fungi, in conjunction with the oxidation and hydrolysis of cellulose that result in energetic losses of the woody fuels during storage. The initial release of heat is mainly produced by the respiration of the live cells of the parenchyma and the bacterial growth. At temperatures above 40°C, the cells that were alive die gradually; the evolution of heat is mainly caused by fungal respiration, yet above 45°C–50°C the chemical reactions releasing heat become increasingly important [18].

The temperature reached in piles of woody residues also depends on the environmental temperature, precipitations, size and compaction of the pile, and the amount and distribution of bark and fine material. In the central area of biomass piles, temperature rises quickly in the first weeks of storage, after which it stabilizes and eventually decreases progressively [17].

Under certain conditions, self-ignition of the piles may occur. This dangerous phenomenon is more likely in big piles, starting in the cavities containing fine elements and bark. The self-combustion risk increases when residues contain large amounts of bark, or when ice layers or fine material accumulations obstruct the cavities, thereby preventing the exchange of heat in the atmosphere.

2.1.2 Drying dynamics in biomass piles

Biomass pile dynamics involve the so-called "chimney effect" (Fig. 4): As the air penetrates inside, the piles form the surface, decreasing the temperature of the

Figure 4: Air flow inside biomass piles.

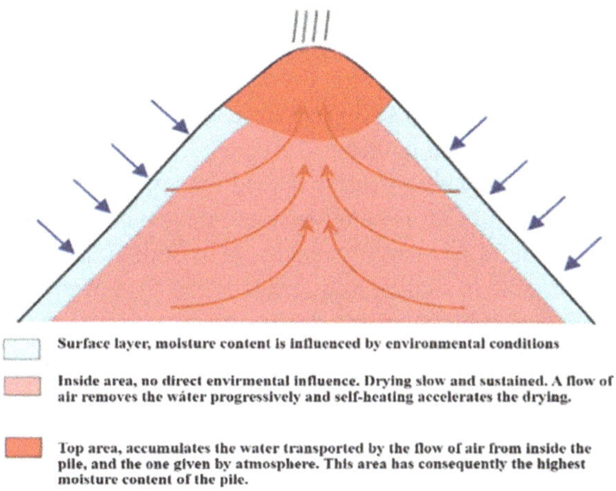

Figure 5: Zones that can be differentiated within a forest biomass pile.

biomass residues. This current of air heats up and acquires the moisture of the residues as it advances towards the pile centre, resulting in a flow of warm and wet air from the centre of the piles to the top. The water consequently flows vertically and accumulates at the pile top, where a strong increase in temperature and vapour exchange to the atmosphere occurs. The increase in temperature is favoured by the accumulation of microorganisms transported by the air flow, which accelerates the growth of the microbiological population and, consequently, the risk of thermogenesis. This vapour loss gives an impression of smoke coming out from the top of piles, hence the name of this phenomenon, *chimney effect.* In a biomass pile, the following three zones can be clearly distinguished (Fig. 5):

- A surface area of a varying depth susceptible to important modifications in the moisture content, due to the direct influence of environmental conditions.
- An inside area where a slow and sustained drying occurs, with no direct influence by the environmental conditions. Moisture loss is instead driven by an air flow that transports water progressively, and self-eating processes that accelerate the drying until moisture content is stabilized.
- A top area where the water transported from the centre and the water from the environment accumulate, resulting in the highest moisture content of the biomass piles.

2.1.3 Experiences with natural air drying of forest chips

The study by Ortiz [17] on the natural air drying of forest wood chips and other forest residues arrived at the following practical conclusions for managing residual biomass piles to be dried:

- The drying season is the factor with the highest influence on the process. Under the climate conditions of central Spain, summer drying is the most favourable. In this season, 2–3 weeks sufficed to reach moisture contents of 20%–30% w.b., whereas in spring 10–15 weeks were needed to reach those values.
- When chips are generated in autumn, even after drying biomass in piles previous to chipping, values of 20%–30% were not reached; a balanced moisture content of 35% was attained within 1–2 weeks.
- Large size chips are capable of faster drying than smaller particle size chips, because the ventilation of air is better between the cavities of bigger chips.
- During storage, microbial activity results in monthly dry matter losses of around 1%. This value is significantly lower in the case of air drying previous to chipping.
- This procedure makes it possible to harvest 60%–75% of relatively dry chips, even in short time periods.

It is important to establish a management of the piles that allows a partial dehumidification without reaching limits that result in important losses of energy and can even be hazardous. Ortiz [17] established consequently the following storage recommendations:

- Piling of chips with conical shape of the piles in winter, with slopes 1:1.
- Piling of chips with trapezoidal shape of the piles in summer.
- Pile sizes should not be larger than 30–50 m^3.
- Make piles by gravity, avoiding compaction.
- Avoid compaction of piles by machinery.
- Prevent as much as possible the presence of fines on the pile surface.
- When vapour is seen coming out of the chip piles, cut the head of the pile.
- When temperatures higher than 60°C are observed, turn over the piles.
- During the dry season, extend the chips in layers of 30 cm and turn over the chips.

2.1.4 Experiences with natural drying of forest bundles under open air, roof covered and controlled temperatures

The study of Dopazo *et al.* [19] monitored the moisture loss of *Eucalyptus camaldulensis* bundles from the Spanish Energy and Pulp Company – ENCE – at the Faculty of Forest Engineering, University of Vigo (Fig. 6).

Weight loss was monitored for 40 days for bundles drying under open air, roof cover and under controlled moisture and temperature conditions in two drying chambers with controlled temperature and relative humidity conditions. Controlled drying chamber 1 simulated mild summer conditions of NW Spain (temperature: 25°C, relative humidity: 55%), whereas chamber 2 monitored the weight loss under warm summer conditions (temperature: 30°C, relative humidity: 40%).

Sharper moisture loss was observed under warm summer conditions. A moisture equilibrium was reached after 1–2 weeks of drying, both in the open air and under the simulated warm summer conditions, to dry camera bundles (Fig. 6). Bundles drying indoor followed the mild summer conditions drying curve (Fig. 6).

2.2 Artificial drying

When it is not possible to decrease the moisture content of the biomass with natural air drying, or when low moisture content values are required (e.g. <10% wet basis, as required for pellet production), artificial drying or a combination of both the methods may be utilized. Artificial drying entails pneumatic or rotatory driers.

- Pneumatic drying is based on the transport of residues with a thermal flux that extracts the moisture of biomass fuels during transport. It is generally utilized when the product has a fine size, or low dehydration is required. Pneumatic driers feature a heat source where conventional fuels of part of the biomass (previously dried) are combusted, a drying channel and a suction system with a cyclonic aspirator.

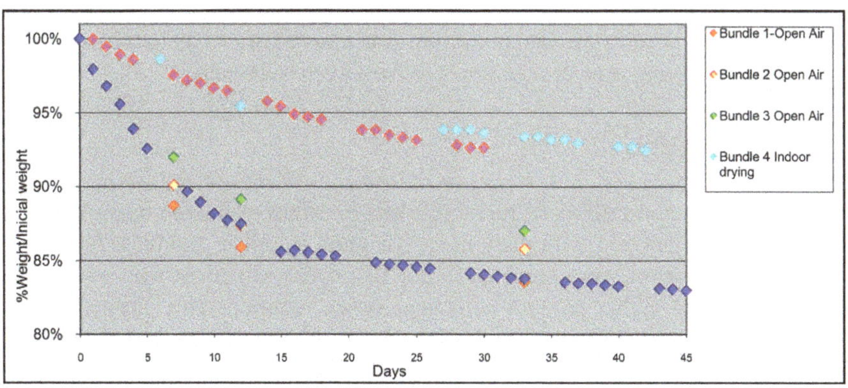

Figure 6: Drying curves for *Eucalyptus* bundles under different conditions.

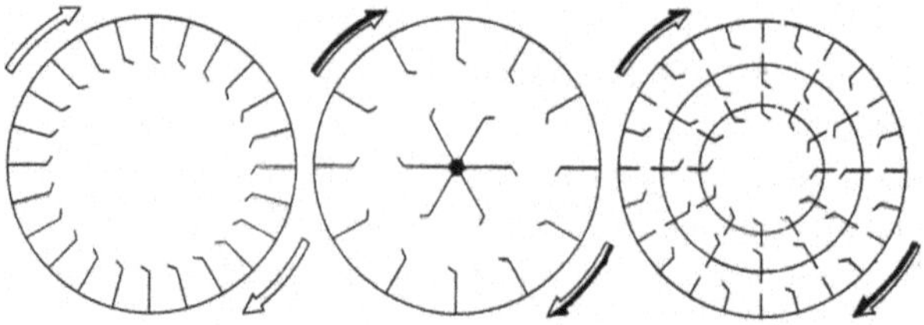

Figure 7: Examples of rotatory dryer designs.

- Rotatory driers are generally utilized when very wet or coarse materials are utilized. Even though they involve the same components as pneumatic driers, in this equipment the channel of circulation is a cylinder of varying length and section. It rotates at a variable speed, inducing close contact between the biomass fuels and the drying flux. Together, the inner slope and the rotation produce an advance of the fuels at a controllable speed. To decrease the length of the rotatory driers, they may be constructed with a single, double or triple pass (Fig. 7); the fuels thus pass through the dryer at alternating routes.

Table 5 summarizes the results obtained from the artificial drying of pine, *Eucalyptus* and Goarse (*Ulex* spp.) in the experimental rotatory dryer (IER-CIEMAT) shown in Fig. 8.

2.2.1 Artificial drying curves for forest residues with varying temperature

Ortiz *et al.* [20] performed a study of controlled artificial drying of thick branches (2.5–5 cm diameter) of *Pinus pinaster* and *Eucalyptus globulus*, with controlled drying at 30°C, 60°C and 105°C in a furnace for 200 h. A sharp decrease in moisture was observed for 60°C and 105°C drying, with equilibrium values reaching within the first 24 and 36 h, respectively, for the two species. At 30°C, which might be representative of the air drying under summer conditions of a thin layer of forest residues, moisture in equilibrium was reached after 100 h, this being the characteristic drying time of medium-size fuels such as branches [21].

3 Size reduction

The grinding and/or milling of forest residues becomes necessary when utilizing biomass in conjunction with equipment designed to utilize products finer than chips. It is, moreover, a pre-requisite for the fabrication of pellets, where a smaller particle size and higher product homogeneity are required. This physical transformation is achieved after the reduction of moisture content in order to obtain smaller particles.

Energy consumption with respect to size reduction depends on the size of particles. As illustrated in Fig. 9, higher consumptions imply more finely sized fuels.

Table 5: Parameters in the controlled drying of several forest species in an experimental rotatory dryer at the CIEMAT Renewable Energy Centre in Spain [18].

Parameter		Pinus	Eucalyptus	Ulex
Capacity	kg/h	1,280	1,292	1,520
Initial moisture	% w.b.	38.8	23.7	59.6
Final moisture	% w.b.	6.3	6.6	4.2
Environmental temperature	°C	26.4	19.4	14.2
Relative humidity environment	%	34.7	45.5	64
Temperature of furnace	°C	438	458	550
Gas temperature entrance	°C	385	410	493
Gas temperature exit	°C	66.8	68.6	64.3
Moisture gas exit	%	12.7	10.67	12.1
Energy consumption	kW	40	63	64
Trommel speed	rpm	9	9	10
Gases	kg/h	7,004	3,369	9,974
Gas flow	Nm^3	5,448	2,620	7,758
Evaporated water	kg/h	443.9	236.5	879
Air/wood (d.b.)	kg/kg	8.94	3.42	16.2
Dry air/wood (d.g.)	kg/kg	0.27	0.2	0.29
Dry air/evaporated water	kg/kg	15.78	14.24	11.35
Consumption of fuel (pellet A1)	kg/h	203	183	187

Figure 8: Experimental rotatory dryer (IER-CIEMAT) [18].

Once the biomass has been reduced, it is necessary to ensure a homogeneous particle size for automatic feeding systems of pellet mills or chip boilers. In addition, fine elements increase ash content and may result in increased ash slagging. The European standard EN 15149-1 [23] establishes limits for both coarse and fine

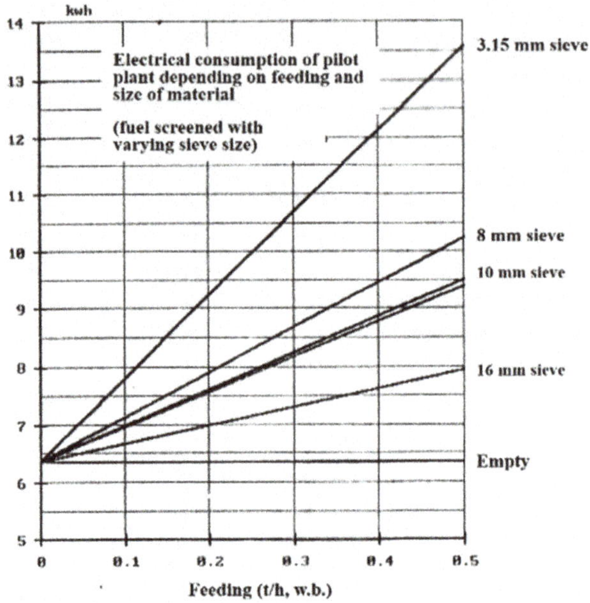

Figure 9: Energy consumption in terms of size reduction at the forest biomass densification pilot plant (shown in Fig. 10) of the Faculty of Forest Engineering, University of Vigo, testing varying fuel feeding amount and particle sizes [18].

Figure 10: Details of the forest biomass densification pilot plant of the Faculty of Forest Engineering, University of Vigo, with a detail of the patented milling hammer system [22]. More information of the pilot plant can be found at: http://lortiz.webs.uvigo.es/?page_id=20

materials, and categorizes chip sizes from P16 to P45 (large chips). Recent ISO standards also include categories for larger chips such as P90 (extra coarse chip). This is the reason behind the growing number of companies performing chip screening at biomass logistic centres (Figs. 11–13) to deliver a homogeneous quality fuel of normalized physical properties.

Figure 11: Screening of residual forest biomass at the Forest biomass pilot plant of the Faculty of Forest Engineering, University of Vigo [18].

Figure 12: Biomass logistic centre in Palas de Reis, NW Spain [24]. Top figure: woody biomass chipper and debarker. Bottom figure: a screen with variable opening size is utilized for controlling the granulometric distribution of wood chips to a P31 or P45 size category. Fine fractions of <8 or 16 mm are separated for pellet production.

Figure 13: Biomass logistic center in Orense, NW Spain [24]. Woody biomass is chipped after storage and dried under open air conditions. Chipped wood and bark are screened for the production of wood chips of normalized sizes P16–P90.

References

[1] Sanz, F. & Piñeiro, G., Aprovechamiento de la biomasa forestal producida por la Cadena Monte-Industria. Parte I: Situación actual y evaluación de sistemas de tratamiento. *CIS-Madera*, **10**, pp. 6–25, 2003.

[2] Pedras, F., Xestión e mecanización da biomasa forestal. Exp. en Galicia, 2008.

[3] IDAE, *Biomasa: Cultivos Energéticos*. IDAE: Madrid, Spain, 2007.

[4] Tolosana, E., Ambrosio, Y., Laina, R., Martinez, R. & Pinillos, F., Rendimientos y costes de diferentes aprovechamientos de la biomasa forestal: Las experiencias en curso en Castilla y León, 2007.

[5] Almagro, C., Tolosana, E., Ambrosio, Y.B.F. & Vignote, S., El estado actual de la gestión de los residuos forestales en España, 2002.

[6] Tolosana, E., Ambrosio, Y., Laina, R. & Martinez, R., Sistemas de aprovechamiento de biomasa en Castilla y León. Las experiencias en curso. *CIDEU*, **5**, pp. 97–106, 2008.

[7] Kauter, D., Lewandowski, I. & Claupein, W., Quantity and quality of harvestable biomass from Populus short rotation coppice for solid fuel use – a review of the physiological basis and management influences. *Biomass and Bioenergy*, **24**, pp. 411–427, 2003.

[8] Larsson, S., Bullard, M.J., Christian, D.G., Knight, J.D., Lainsbury, M.A. & Parker, S.R. Commercial varieties from the Swedish willow breeding

programme. Biomass Energy Crop. II, University of York, York, UK, 18–21 December 2001, Association of Applied Biologists, 2001, pp. 193–198.

[9] Sixto, H., Hernandez, M.J., Barrio, M., Carrasco, J. & Cañellas, I., Plantaciones del género Populus para la producción de biomasa con fines energéticos: revision. *Sist Recur For*, **16**, pp. 277–294, 2007.

[10] Vega-Nieva, D., Dopazo, R. & Ortiz, L., Reviewing the potential of forest bioenergy plantations: woody energy crop plantations management and breeding for increasing biomass productivity. *World Bioenergy*, 2008.

[11] Vega-Nieva, D., Dopazo, R. & Ortiz, L., Experiencias con cultivos energéticos para producir biocombustibles sólidos en el mundo. *3er Congr. Int. Bioenergía*, 2008.

[12] Ciria, M.P., Mazon, P., Carrasco, J.E. & Fernandez, J., Effect of rotation age on the productivity of poplar grown at high plantation density. *8th Eur. Biomass Conf.*, 1995, pp. 489–494.

[13] Ciria, M.P., Gonzalez, E., Mazon, P. & Carrasco, J.E., Influence of the rotation age and plant density on the composition and quality of poplar biomass. *9th Eur. Bioenergy Conf. 1st Eur. Energy from Biomass Technol. Exhib.*, 1996, pp. 962–973.

[14] Ciria, M.P., La Producción de Chopo con fines energéticos en España. I Congr. Int. Bioenergía. La Bioenergía, una oportunidad, una necesidad, 2006.

[15] San Miguel, A. & Montota, J.M., Resultados de los primeros 5 años de producción de tallares de chopo en rotación corta (2–5 años). *An INIA, Ser For*, **8**, pp. 73–91, 1984.

[16] Gasol, C.M., Martínez, S., Rigola, M., Rieradevall, J., Anton, A, Carrasco, J., *et al.*, Feasibility assessment of poplar bioenergy systems in the Southern Europe. *Renew Sustain Energy Rev*, **13**, pp. 801–812, 2009.

[17] Ortiz, L., *Air Drying of Forest Residual Biomass*. Polytechnic School of Madrid: Madrid, 1989.

[18] Ortiz, L., Aprovechamiento Energetico de la Biomasa Forestal, 1996.

[19] Dopazo, R., Vega-Nieva, D. & Ortiz, L., Drying of Eucalyptus bundles under open air, roof-covered and under simulated mild and warm summer conditions, 2011.

[20] Ortiz, L., Tejada, A., Vazquez, A. & Piñeiro, G., Aprovechamiento de la Biomasa Forestal producida por la Cadena Monte-Industria. Parte III: Producción de elementos densificados. *CIS-Madera*, **10**, pp. 38–75, 2003.

[21] Anderson, H.E., Aids to determining fuel models for estimating fire behavior, 1982.

[22] Ortiz, L., La biomasa como fuente de energía renovable, 2006.

[23] EN15149-1, Solid biofuels. Determination of particle size distribution. Part 1: Oscillating screen method using sieve apertures of 1 mm and above, 2010.

[24] Vega-Nieva, D., Alvarez, C. & Ortiz, L., Results of new laboratory methods and slagging classification systems for the prediction and quantification of ash slagging in woody and herbaceous biomass fuels. *Cent. Eur. Biomass Conf.*, 2014.

CHAPTER 4

Biomass Pelletization Process

Angela Garcia-Maraver[1] & Manuel Carpio[2]
[1]*Department of Civil Engineering, University of Granada, Campus de Fuente Nueva, 18071 Granada, Spain.*
[2]*Department of Building Construction, Campus de Fuente Nueva, 18071 Granada, Spain.*

Abstract

The production of energy by means of biomass has shown a clear trend towards the use of pellets due to their homogeneous size, which facilitates handling and feeding while also reducing costs associated with storage and transportation. The pellet quality depends on the properties of the feedstock and on the operative variables of the densification process. Quality parameters can be modified by adapting the process to the raw material to be pelletized. Whereas wood pellets from forestry residues already have successfully established technologies and markets, there is a need to develop the pelletization of agricultural biomass, which is periodically planted and harvested, and holds great energy potential, especially in rural areas.
Keywords: Pellet, manufacturing conditions, pelletization stages, biomass.

1 Introduction

In recent years, the production of thermal energy by means of biomass on a small scale has shown a clear trend towards densified biofuels (pellets) [1]. The homogeneous size facilitates an automatic or semi-automatic treatment and, thus, resolves the disadvantages of the traditional domestic use of biomass [2,3]. The use of densified biofuels also reduces the costs associated with handling, storage and transportation.

Despite the vast generation of agricultural and forestry residues, their utilization as fuel is still low in light of their energy potential. Some important factors that influence the level of energy usage of these residues are their low bulk density [4,5], local availability, the fact that they are dispersed over a relatively large area and a lack of information concerning fuel feeding, combustion and emission

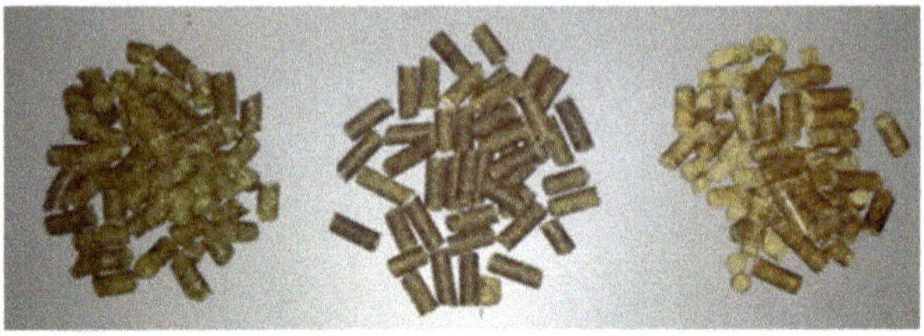

Figure 1: Pellets made from different types of residual biomass under different pelletization conditions.

characteristics. This information is important for the design and efficient operation of combustion systems fully adapted to the biomass of each region.

The management of these residues is hampered by the costs associated with treatments needed for their proper removal [3]. Pelleting residual biomass would mean the conversion of a residue difficult to manage into an energy resource most adequate for use in the areas where each grove exists.

The final quality of pellets varies depending on the raw material properties and the manufacturing process [6] (Fig. 1). Although the inorganic and organic components of the different raw materials cannot be modified, certain variables dependent on the pelletization process can be controlled to optimize the production efficiency and enhance the quality of the finished product [5].

The aim of this chapter is to analyse the operative variables of the pelletization process that can be adapted to each type of biomass to improve the characteristics of the final pellets produced, according to the quality guidelines for biomass established by the European Standards EN14961-2 [7] and EN14961-6 [8].

2 The Pelletization Process

The biomass pelletization process consists of multiple steps including raw material pre-treatment, pelletization and post-treatment (Fig. 2).

2.1 Feedstock storage

An effective feedstock storage system is necessary so as to keep biomass away from impurities and offer adequate protection from rain and moisture in general. Rain may actually damage the feedstock by increasing its moisture content to such a high level that the drying process becomes unviable.

2.2 Removal of undesirable impurities

Raw material should be filtered before grinding to remove materials such as stones or metal fragments [9].

Many pellet plants are equipped with systems separating virgin biomass from inert materials such as stones and soil, or magnetic separators for metal impurity

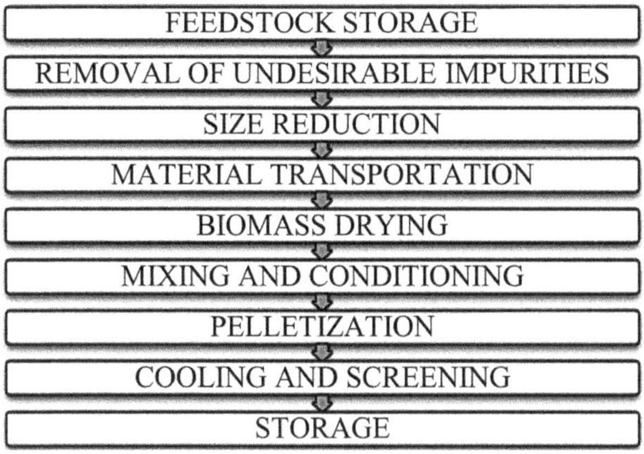

Figure 2: Stages of the pelletization process.

removal. The presence of even the slightest amount of such materials in the final product is unacceptable. At the same time, they can cause serious damage to the mechanical equipment, i.e. in the hammer mill or in the pellet press.

2.3 Size reduction

The raw material must be reduced to a uniform size that is adequate for the pellet mill. The milled material going into the pellet mill has to be smaller than the die holes to prevent blocking of the holes. Therefore, materials should be ground to a size no bigger than the diameter of the pellet (~6 mm), producing a substance with a consistency similar to bread crumbs [10].

For pelletizing, it is generally assumed that small particles with a large surface area will increase density and result in stronger pellets [11–13]. A narrow particle size distribution of the raw material also facilitates even moisture distribution during the drying stage. In a mixture of different sizes, small particles might become overly dried in the process and make self-bonding more difficult later on in the densification stage [14]. A mixture of particle sizes is considered beneficial, because this increases the durability of the pellets [15].

The size reduction equipment will vary depending on the characteristics of the raw materials. Chippers and shredders are used for raw materials with large diameters, while a hammer mill is more appropriate for chips or herbaceous raw materials.

In addition, the diameter of the raw material will influence the choice of equipment [16], chippers/shredders being adequate for raw materials with a diameter over 2.54 cm. Hammer mills are normally the most suitable for size reduction; however, there is a limitation on the size of the input material, which has to have a diameter of <2.54 cm to be processed.

Generally, chippers/shredders are used as a first step. Then, once the size of the chipped material has been reduced, the hammer mill is used as a finishing mill.

2.4 Material transportation

Once reduction has taken place, the material must be moved on to the next process. Screw conveyors are the most commonly used systems for transporting the material through the pelletization process because of the low cost involved.

Alternative options would include the use of a fan followed by a cyclone separator that separates air from the milled material. The air leaves the top of the cyclone, and the milled material is thrown to the outside edge of the cone and then falls out below [16].

2.5 Biomass drying

In pelletization techniques, the moisture percentage of the raw material should be between 10% and 20% to assure high-quality pellet production [16] (Fig. 3). However, the exact percentage to produce quality pellets is specific to each raw material [17].

Drying solid biofuels is a key factor, since wet raw materials result in low combustion temperatures, low energy efficiency and high emissions of hydrocarbons and particles [18]. In addition, it is very important to choose the right dryer, which may constitute the largest capital expenditure in a pellet production plant [16] and can dramatically increase the production cost per tonne [9,10].

Though fossil fuels can be used as the heat source of the dryers, they would increase pellet production costs while reducing the environmental friendly credentials of the biomass fuel pellets. Hence, the most common and cheapest solution is to generate heat by burning the pellets produced in situ, or the dried chips used for making the pellets. Yet, it is important to remember that the fly ash contained in the flue gas used in directly heated drum dryers remains in the dried sawdust, which leads to a higher ash content of the pellets and a certain degree of contamination by heavy metals [19].

Dryers can be classified depending on the medium used. According to Mujumdar [20], superheated steam dryers have some key advantages compared with air dryers – such as lumber or rotary dryers – because no oxidation or combustion reactions

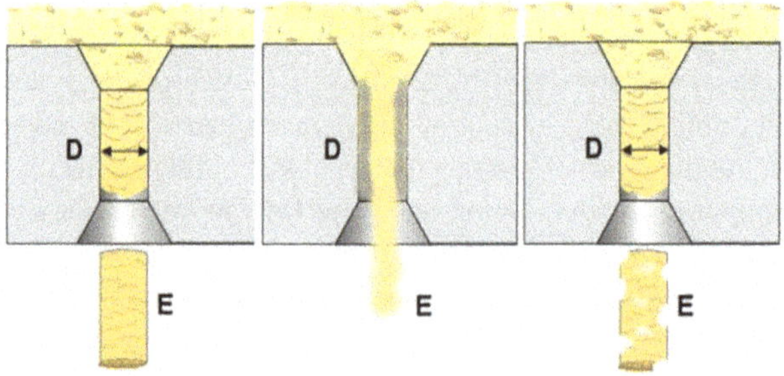

Figure 3: Influence of the moisture content of raw material in pellet production [16].

are possible. Steam dryers have higher drying rates than air and gas dryers, and do not entail any danger of fire or explosions, while allowing toxic or valuable liquids to be separated in condensers [18]. These systems are more complex, however, and even minor steam leakage is devastating in terms of the energy efficiency of the steam dryer [21]. The most commonly used systems are the flue gas rotary dryers, which combine high inlet temperature with long residence time, and thus can result in pyrolysis and partial gasification [22].

2.6 Mixing and conditioning

Not all raw materials require mixing. Nevertheless, if necessary, mixers are used after the raw material has been milled and dried to: (1) get a more consistent material blend to be fed into the pellet mill when the raw material presents significant changes in moisture percentage, binding properties or material density [16] or (2) produce a homogeneous mixture of raw materials in the case of pellets made up of different raw materials.

Once the raw material has been mixed and its consistency is high enough, it may require the addition of other constituents able to increase the productivity of the pellet mill and enhance the final properties of the pellets produced.

Additives play a major role in wood pellet characteristics. They are a subject of major interest: as binding agents for the biomass raw material, they serve to improve pellet durability and physical quality, reduce the dust, improve pelleting efficiency and reduce energy costs.

A maximum content of 2% of additives is permitted in woody pellets [7]. No limitation exists for the non-woody pellets [8], though it is required to indicate the type and quantity used. To produce wood pellets with desired physical and thermal characteristics, the additive should be suited to the right biomass material. The most common additives are: (1) water, which is used if the moisture content of the mixed material is too low and (2) binders, which act as a glue between the particles if the lignin content of the material is not enough to hold a pellet together. Lignin is a natural, optimal binder of biomass, because it melts under the heat of the pellet mill [23]. If the lignin content of the biomass is low, it may be necessary to add other additives, however. One of the simplest binders is vegetable oil, which also aids the pellet process by reducing the frequency of blocked dies [16], but the most widely used substance overall is starch [24].

As concluded by Tarasov *et al.* [24] and summarized below, each additive will produce unique physical and thermal characteristics when used with different biomass materials:

- Lignosulphonate additives result in the best mechanical durability values for wood pellets but do not display high particle density. As they do not affect the calorific value of the wood pellets, they significantly increase sodium and sulphur content, and consequently increase emissions.
- Starch additives are the most widely used because they reduce the final moisture content more than lignosulphonate additives. However, too much starch

will make the final product extremely dry, which affects the mechanical durability of the wood pellets.
- Additives such as motor oil, corn starch, sodium carbonate, urea, vegetable oil and dolomite decrease the wood pellet particle density and contribute to the pellet process by reducing the frequency of blocked dies [16].

Corn starch and dolomite additives are the most effective in reducing the wood pellet particle density. All types of starch (native wheat starch, oxidized corn starch, native potato starch and oxidized potato starch) increase the mechanical durability of the wood pellets, the best results for mechanical durability being obtained by adding oxidized corn starch. Motor and vegetable oil additives increase the calorific value minimally, while corn starch and dolomite additives reduce the calorific values of wood pellets. Wheat starch is an additive that significantly reduces ash formation, but dolomite additive increases ash formation as well as the ash melting point in wood pellet combustion. Both corn starch and dolomite additives significantly increase carbon monoxide emissions.

Many pelletizing machines come with a built-in steam conditioning chamber [10]. Super-heated steam, at temperatures above 100°C, is used to soften the wood before it is compacted. Steam conditioning is not necessary, but it does make the raw material less abrasive to the pelletizing equipment and this helps reduce the maintenance costs [9].

2.7 Pellet production

Pellets are finally made by pellet mills, also known as pellet presses or extruders, which are available in a range of sizes. Generally, every 100 horsepower provides a capacity of approximately 1 ton of pellets per hour [10].

Pellets are created by using a great deal of pressure to force the raw material through holes in the die. The pellet mills are mainly formed by two elements: the die with the holes that act as the mould and the rollers that force the raw materials to go through the holes of the die.

As the pressure and friction increase, the temperature of the wood also increases. This allows the lignin to soften and the fibre to be reshaped into pellet form [9]. Simultaneously, the moisture content is reduced because of the higher temperature.

2.7.1 Types of pellet mills

According to the shape of the die, pellet mills can be classified as: (1) flat die pellet mills, which are used for small- to medium-scale pellet industries and (2) round die pellet mills, which are applied to medium- and large-scale pellet industries [16].

Flat die mills use a flat die with slots (Fig. 4). The powder is introduced into the top of the die and, as the die rotates, a roller presses the powder through the holes in the die. A cutter on the other side of the die cuts the exposed pellet free [25].

In the round/ring die, there are radial slots throughout the die. The die is positioned vertically, powder is fed into the centre of the die and spreaders evenly

Figure 4: Flat pellet mill with two rollers.

distribute the powder. Two or more rollers then compress the powder through the die holes. Two cutters are used to cut the pellets free from the outside of the die [26].

In both cases, there has to be a rotary element, which can be the die or the rollers. In the case of flat dies, the rollers are the moving element, whereas the most common design of ring dies is where the die is powered and rotating, and the rollers move due to the friction and movement of the die.

The advantages and disadvantages of each type of pellet mill are summarized in Table 1.

2.7.2 Types of die

There are different die types that vary with regard to hole depth, hole pitch, hole width and die material [16].

- The depth of the hole is not related to the length of the pellet. It establishes the time the material is exposed to heat and pressure. Longer holes are recommended for materials with low lignin content to achieve better melting and, therefore, a harder, denser pellet.
- The pitch angle of the hole influences pellet compression, and it should be designed according to the feedstock.
- For biomass pellets, the most common diameters are 6 and 8 mm. These diameters, as well as the length of the pellets, are limited by the European Standards

Table 1: Advantages and disadvantages of flat and round pellet mills [16].

	Flat die	Round die
Advantages	Easier to clean Quick access to the chamber Compact design	Lower costs of roller and die consumables
	Visibility of the pellets produced Robust for problematic feedstock (wider tolerance)	Extra friction resulting in more heat and better quality of the pellets
Disadvantages	Possible uneven roller and die wearing	Extra friction resulting in more energy consumption Large size and weight Difficult access to rollers and dies
	Slipping action of the rollers (can be solved by tapering the rollers)	Manual roller adjustment (can be solved by remote access) High costs No visibility of the pellet process

[7,8] to assure that the pellets can be fed by the screw conveyors of standardized stoves and boilers without blocking the feeding systems. Experience has shown that thinner pellets allow a more uniform combustion rate than thicker ones, especially in small furnaces [11]. The length of the pellets affects the fuel feeding properties: the shorter the pellets, the easier to arrange for continuous flow [11].

- Pellet mill dies are made from several different types of metal. Dies must be resistant to abrasion, strong enough to handle the forces of pellet production and corrosion resistant without negatively affecting productivity. The materials most often used are carbon steel alloy, stainless steel alloy and high chrome alloy. Obviously, the most expensive materials ensure higher resistances and productivities.

These parameters have to be optimized in view of the particular feedstock to guarantee good productivity and to avoid blocking the dies. Blocked dies can be found when: (1) the raw material generates more resistant force while passing through the die than the force generated by the roller and/or (2) the die hole is corroded [16].

If the die gets blocked, the material has to be drilled out. To facilitate this process, it is best to work when the die is still warm, previous to the formation of hard pellets within the die holes.

2.8 Cooling and screening

Moisture resistance is an important pellet property. The pelletizing process implies high temperatures, and attention must be paid to proper cooling and heat removal before the pellets leave a production plant, especially with regard to the storage stage [12,27]. If moisture content exceeds 20% of d.b., bacterial growth might occur [28], causing material degradation and self-heating, which in the worst case scenario might result in self-ignition [27]. In sum, the cooling process is critical for pellet strength and durability. When pellets leave the extruder, they are hot (90°C–95°C) and soft [10]; then, they are gradually air cooled, which allows the lignin to solidify and strengthen the pellets. In contrast to the drying process, cooling does not involve the addition of energy. There are three types of coolers: vertical, horizontal and continuous flow [9].

Once pellets have cooled, they pass over a vibrating screen to remove any fine material [10]. These 'fines' are augured back into the pelletizing process to ensure that no raw material is wasted. Screening ensures that the fuel source is clean and as near to dust free as possible. Once screened, pellets are ready to be packaged for the desired end use [9].

3 European Guidelines

According to the quality parameters described in a previous chapter, and considering the specific characteristics of woody pellets from forestry residues and non-woody biomass from agricultural residues, the following European guidelines have been established (Tables 2–4).

3.1 European guidelines for woody pellets

The EN 14961-2 [7] defines three quality categories (A1, A2 and B) for woody pellets to guarantee their proper use and resistance (Table 2).

3.2 European guidelines for non-woody pellets or pellets from mixtures of biomasses

The EN 14961-6 [8] defines specific categories for straw, miscanthus and reed canary grass pellets (Table 3) in addition to two quality categories (A and B class pellets, produced from herbaceous and fruit biomass and blends and mixtures that can also include woody biomass) (Table 4). These guidelines are generally less restrictive than those established for woody pellets in EN 14961-2 [7].

4 State-of-the-Art of the Pelletization of Forestry and Agricultural Residues

Pelletizing technologies for the production of upgraded biomass fuels must be improved to lower costs and enhance fuel quality. Improvement may entail the

Table 2: European normative guidelines for woody pellets [7].

	Units	A1	A2	B
Diameter and length (D and L)	mm	D06: D ≤ 6 ± 1 and 3.15 ≤ L ≤ 40 D08: D ≤ 8 ± 1 and 3.15 ≤ L ≤ 40		
Moisture, M	% as received	M10: ≤10		
Ash, A	% dry basis	A0.7 ≤ 0.7	A1.5 ≤ 1.5	A3.0 ≤ 3.0
Mechanical durability, DU	% as received	DU97.5 ≥ 97.5		DU96.5 ≥ 96.5
Fines	% as received	F1.0 ≤ 1.0		
Additives	% dry basis	≤2% (type and quantity to be specified)		
Lower heating value as received, Q	MJ/kg	Q16.5: 16.5 ≤ Q ≤ 19	Q16.3: 16.3 ≤ Q ≤ 19	Q16.0: 16.0 ≤ Q ≤ 19
Bulk density, BD	kg/m^3 as received	BD600 ≥ 600		
Nitrogen, N	% dry basis	N0.3 ≤ 0.3	N0.5 ≤ 0.5	N1.0 ≤ 1.0
Sulphur, S	% dry basis	S0.03 ≤ 0.03		S0.04 ≤ 0.04
Chlorine, Cl	% dry basis	Cl0.02 ≤ 0.02		Cl0.03 ≤ 0.03

Table 3: European normative guidelines for straw, miscanthus and reed canary grass pellets [8].

	Units	Straw	Miscanthus	Reed canary grass
Diameter and length, D and L	mm	D06-10: D ± 1; 3.15 ≤ L ≤ 40 D12-25: D ± 1; 3.15 ≤ L ≤ 50		
Moisture, M	% as received	M10 ≤ 10		M12 ≤ 12
Ash, A	% dry basis	A6.0 ≤ 6 A6.0+ >6	A4.0 ≤ 4	A8.0 ≤ 8 A8.0+ > 8
Mechanical durability, DU	% as received	DU97.5 ≥ 97.5		DU96.5 ≥ 96.5
Fines	% as received	F1.0 ≤ 1.0		
Additives	% dry basis	Type and quantity		
Lower heating value as received, Q	MJ/kg	Minimum value		Q14.5 ≥ 14.5
Bulk density, BD	kg/m^3 as received	BD600 ≥ 600	BD580 ≥ 580	BD550 ≥ 550
Nitrogen, N	% dry basis	N0.7 ≤ 0.7	N0.5 ≤ 0.5	N2.0 ≤ 2.0
Sulphur, S	% dry basis	S0.10 ≤ 0.10	S0.05 ≤ 0.05	S0.20 ≤ 0.20
Chlorine, Cl	% dry basis	Cl0.10 ≤ 0.10	Cl0.8 ≤ 0.8	Cl0.10 ≤ 0.10

Table 4: European normative guidelines for pellets produced from herbaceous and fruit biomass and blends and mixtures [8].

	Units	A	B
Diameter and length, D and L	mm	D06-10: D ± 1; 3.15 ≤ L ≤ 40 D12-25: D ± 1; 3.15 ≤ L ≤ 50	
Moisture, M	% as received	M12 ≤ 12	M15 ≤ 15
Ash, A	% dry basis	A5.0 ≤ 5	A10 ≤ 10
Mechanical durability, DU	% as received	DU97.5 ≥ 97.5	DU96.5 ≥ 96.5
Fines	% as received	F2.0 ≤ 2.0	F3.0 ≤ 3.0
Additives	% dry basis	Type and quantity	
Lower heating value as received, Q	MJ/kg	Q14.1 ≥ 14.1	Q13.2 ≥ 13.2
Bulk density, BD	kg/m^3 as received	BD600 ≥ 600	
Nitrogen, N	% dry basis	N1.5 ≤ 1.5	N2.0 ≤ 2.0
Sulphur, S	% dry basis	S0.20 ≤ 0.20	
Chlorine, Cl	% dry basis	Cl0.20 ≤ 0.20	Cl0.30 ≤ 0.30

proper selection of matrices, testing and evaluation of bio-additives for quality improvement and reduction of operating costs, and development and testing of pre-treatment technologies for proper conditioning of the raw material. Moreover, future pellet production will have to cope with raw materials other than those most commonly used at present, i.e. wood shavings and sawdust. Due to steadily growing pellet markets in Europe and worldwide, the production of pellets from woodchips, forestry residues, short rotation coppice and different kinds of herbaceous biomass fuels will increase in the future, which makes respective R&D activities necessary [29].

As stated by Carroll and Finnan [30], the shortage of biomass from forestry production has increased the use of a wide range of alternative biomass feedstock to generate renewable energy [31]. This view is shared by Verma *et al.* [32,33], who state that 'the limited availability of woody biomass for energetic purposes leads to the introduction of agro-forestry driven residues'. While an increasing proportion of the biomass needed for renewable energy is being imported from countries rich in woody biomass such as Sweden or Germany, agricultural residues and energy crops already play a large role in renewable energy generation [31] and are predicted to play a major part in the future energy crop mix [34].

The main problems with agricultural raw materials, in comparison to woody raw materials, are the higher ash content, the lower ash softening temperature and the higher risks of corrosion and fouling [35–37]. Additives can counteract these problems to some degree, but at the same time they increase the costs, and problems with high ash content would still remain. One way to increase the use of these raw materials is to mix them with sawdust/shavings, and to adapt their proportions to the combustion conditions in the boiler (fluidized bed boiler, burner, etc.). Another problem with these materials is that the production costs may be high, especially for cultivated crops, requiring fertilizers, specialist machinery, etc. From a natural resource point of view, it is important that the net energy yield of using pellets as a fuel is as high as possible [38].

Most raw materials also have low bulk densities, resulting in high transport and storage costs. Pelletization affords benefits such as increased bulk density, therefore reducing transportation and storage costs and providing better material feeding with less dust formation [39], if we consider the factors that affect milling properties and the agglomerating character of the raw materials [11].

5 Conclusions

The pelletization process entails several stages that can be modified depending on the type of biomass used. These raw materials generally call for pre-treatment because of their origin and characteristics.

To determine the optimal pelletization conditions for each raw material, the densification process must first be tested under different conditions in terms of: compression force, moisture of the raw material, temperature of the pelletization and percentage of additives, when necessary.

The pellets produced under the conditions analysed will have different values in view of their quality properties, and only those that fulfil the requirements established by the existing norms should be used as fuel.

References

[1] Verma, V.K., Bram, S. & De Ruyck, J., Small scale biomass heating systems: standards, quality labelling and market driving factors – an EU outlook. *Biomass and Bioenergy*, **33**, pp. 1393–1402, 2009.

[2] Holm, J.K., Henriksen, U.B., Hustad, J.E. & Sorensen, L.H., Toward an understanding of controlling parameters in softwood and hardwood pellets production. *Energy & Fuels*, **20**, pp. 2686–2094, 2006.

[3] Miranda, T., Arranz, J.I., Montero, I., Román, S., Rojas, C.V. & Nogales, S., Characterization and combustion of olive pomace and forest residue pellets. *4th International Congress of Energy and Environment Engineering and Management*, **103**, 91–96, 2012.

[4] Werther, J., Saenger, M., Hartge, E.-U., Ogada, T. & Siagi, Z., Combustion of agricultural residues. *Progress in Energy and Combustion Science*, **26**, pp. 1–27, 2000.

[5] Carone, M.T., Pantaleo, A. & Pellerano, A., Influence of process parameters and biomass characteristics on the durability of pellets from the pruning residues of *Olea europaea* L. *Biomass and Bioenergy*, **35**, pp. 402–410, 2011.

[6] Samuelsson, R., Thyrel, M., Sjöström, M. & Lestander, T.A., Effect of biomaterial characteristics on pelletizing properties and biofuel pellet quality. *Fuel Processing Technology*, **90**, pp. 1129–1134, 2009.
[7] EN14961-2, Solid biofuels. Fuel specification and classes. Part 2: wood pellets for non-industrial use, 2011.
[8] EN14961-6, Solid biofuels. Fuel specifications and classes. Part 6: non-woody pellets for non-industrial use, 2012.
[9] Peksa-Blanchard, M., Dolzan, P., Grassi, A., Heinimö Junginger, M.J., Ranta, T., *et al.*, Global wood pellets markets and industry: policy drivers, market status and raw material, Potential IEA Bioenergy Task 40, 2007.
[10] Karkania, V., Fanara, E. & Zabaniotou, A., Review of sustainable biomass pellets production – a study for agricultural residues pellets' market in Greece. *Renewable & Sustainable Energy Reviews*, **16**, pp. 1426–1436, 2012.
[11] Lehtikangas, P., Quality properties of pelletised sawdust, logging residues and bark. *Biomass and Bioenergy*, **20**, pp. 351–360, 2001.
[12] Mani, S., Tabil, L.G. & Sokhansanj, S., Effects of compressive force, particle size and moisture content on mechanical properties of biomass pellets from grasses. *Biomass and Bioenergy*, **30**, pp. 648–654, 2006.
[13] Kaliyan, N. & Vance Morey, R., Factors affecting strength and durability of densified biomass products. *Biomass and Bioenergy*, **33**, pp. 337–359, 2009.
[14] Bergström, D., Israelsson, S., Öhman, M., Dahlqvist, S.-A., Gref, R., Boman, C., *et al.*, Effects of raw material particle size distribution on the characteristics of Scots pine sawdust fuel pellets. *Dimethyl Ether Spec Sect*, **89**, pp. 1324–1329, 2008.
[15] Okuda, N. & Sato, M., Manufacture and mechanical properties of binderless boards from kenaf core. *Journal of Wood Science*, **50**, pp. 53–61, 2004.
[16] PelHeat, Biomass pellet production guide. 2010.
[17] Garcia-Maraver, A., *Optimization of the Pelletization Process of Agricultural Wastes Originating from Olive Farms for Their Application in Domestic Boilers*. University of Granada, 2013.
[18] Ståhl, M., Granström, K., Berghel, J. & Renström, R., Industrial processes for biomass drying and their effects on the quality properties of wood pellets. *Biomass and Bioenergy*, **27**, pp. 621–628, 2004.
[19] Nussbaumer, T., Combustion and co-combustion of biomass: â €‰ fundamentals, technologies, and primary measures for emission reduction. *Energy Fuels*, **17**, pp. 1510–1521, 2003.
[20] Mujumdar, A.S., Superheated steam drying. In *Handbook of Industrial Drying*, ed. A.S. Mujumdar, Marcel Dekker, Inc.: New York, Basel, p. 195, 1995.
[21] Berghel, J. & Renström, R., *Design Criteria, Key Parameters and Performance Results of a Fluidized Dryer*. Department of Environmental and Energy Systems, Karlstad University, 2000.
[22] Wimmerstedt, R. & Hallström, A., *Drying of Peat and Biofuels. Techniques, Economy and Development Needs,* LUTKDH/TKKA-3002. Report, Lund University, Lund, Sweden, 1984.

[23] Garcia-Maraver, A., Salvachúa, D., Martínez, M.J., Diaz, L.F. & Zamorano, M., Analysis of the relation between the cellulose, hemicellulose and lignin content and the thermal behavior of residual biomass from olive trees. *Waste Management*, **33**, 2245–2249, 2013.
[24] Tarasov, D., Shahi, C. & Leitch, M., Effect of additives on wood pellet physical and thermal characteristics: a review. *ISRN For*, **2013**, p. 6, 2013.
[25] Pietsch, W., *Agglomeration in Industry*, Wiley-VCH-Verlag, 2005.
[26] Matz, S.A., *Chemistry and Technology of Cereals as Food and Feed*, Springer, 1991.
[27] Rhén, C., Gref, R., Sjöström, M. & Wästerlund, I., Effects of raw material moisture content, densification pressure and temperature on some properties of Norway spruce pellets. *Fuel Processing and Technology*, **87**, pp. 11–16, 2005.
[28] Lehtikangas, P., Storage effects on pelletised sawdust, logging residues and bark. *Biomass and Bioenergy*, **19**, pp. 287–293, 2000.
[29] Van Loo, S. & Koppejan, J., *The Handbook of Biomass Combustion and Co-firing*. Earthscan: Sterling, VA, 2008.
[30] Carroll, J. & Finnan, J., Emissions and efficiencies from the combustion of agricultural feedstock pellets using a small scale tilting grate boiler. *Biosystems Engineering*, **115**, pp. 50–55, 2013.
[31] Woods, J., Tipper, R., Brown, G., Diaz-Chavez, R., Lovell, J. & de Groot, P., *Evaluating the Sustainability of Co-firing in the UK*, URN 06/196. Department of Trade and Industry, UK, 2006.
[32] Verma, V.K., Bram, S., Gauthier, G. & De Ruyck, J., Evaluation of the performance of a multi-fuel domestic boiler with respect to the existing European standard and quality labels: part 1. *Biomass and Bioenergy*, **35**, pp. 80–89, 2011.
[33] Verma, V.K., Bram, S., Gauthier, G. & De Ruyck, J., Performance of a domestic pellet boiler as a function of operational loads: part 2. *Biomass and Bioenergy*, **35**, pp. 272–279, 2011.
[34] Clifton-Brown, J.C., Breuer, J. & Jones, M.B., Carbon mitigation by the energy crop, *Miscanthus*. *Global Change Biology*, **13**, pp. 2296–2307, 2007.
[35] Jenkins, B.M., Baxter, L.L., Miles, Jr. T.R. & Miles, T.R., Combustion properties of biomass. *Fuel Processing and Technology*, **54**, pp. 17–46, 1998.
[36] Nikolaisen, L., Nørgaard, J.T., Hjuler, K., Busk, J., Junker, H., Sander, B., *et al.*, Quality characteristics of biofuel pellets, Vol. 51161/00-0. Danish Technological Institute, 2002.
[37] Obernberger, I. & Thek, G., Physical characterisation and chemical composition of densified biomass fuels with regard to their combustion behaviour. *Biomass and Bioenergy*, **27**, pp. 653–669, 2004.
[38] Nilsson, D., Bernesson, S. & Hansson, P.-A., Pellet production from agricultural raw materials – a systems study. *Biomass and Bioenergy*, **35**, pp. 679–689, 2011.
[39] Gilbert, P., Ryu, C., Sharifi, V. & Swithenbank, J., Effect of process parameters on pelletisation of herbaceous crops. *Fuel*, **88**, pp. 1491–1497, 2009.

CHAPTER 5

Biomass Pellet-Fired Boilers

Jose Antonio Perez-Jimenez
Department of Biomass, CTAER Andalucia Foundation, Scientific and Technology Park, Jaen, Spain.

Abstract

Wood fire has been used as a heat source for thousands of years. Nonetheless, the methods for burning wood have progressed to very automatic and controlled combustion systems. The pressure to develop systems that minimize air pollution and maximize heating efficiency has led to imaginative and innovative new designs. Provided that firewood is grown in a sustainable manner and used in efficient combustion systems with insignificant hydrocarbon emissions, firewood is a renewable energy source. Already today, it is a significant heating source in most of the world. Due to its potential of being CO_2 neutral, an increased use of small-scale combustion can have a significant impact on reducing greenhouse gas emissions. The fuels in use are mainly wood logs, but the use of densified biomass in the form of pellets or briquettes is playing an important role, especially in the residential and public sectors. In this regard, it has to be noted that the use of traditional batch-fired systems burning firewood has been augmented because of the introduction of systems designed to burn pelletized wood wastes, agricultural grains or woodchips. Due to the fact that intelligent design and sophisticated technology have resulted in several types of biomass heating systems, the aim of this chapter is to provide the reader with an overview of the domestic pellet boilers.
Keywords: Biomass, pellet, boiler, efficiency, stoves, fireplace.

1 Domestic Pellet-Burning Appliances

In spite of its long history, biomass is the most difficult of the commonly used heating fuels to burn clean and efficiently, especially at a small scale. Residential wood-fired heating systems typically have considerable amounts of particulate, carbon monoxide and other unburned gaseous emissions compared with systems fired by natural gas and oil-fired systems. The more recent development of wood-fired appliances has largely focused on the reduction of these undesirable emissions to retain the desirability of wood as a fuel source. The challenge in designing batch-fired heating systems is to provide a controlled rate of heat output over as long duration as possible, while capturing as much of the heat released by the fuel as practicable. Therefore, the development of efficient and clean-burning systems has preoccupied product designers for decades.

The use of wood pellets for domestic heating is a rather recent phenomenon. Pellet-fired appliances were developed in the 1980s but significant market penetration of pellet stoves has only taken place during the past 15 years. With the introduction of wood pellets and automatic combustion systems in recent years, it has become much easier to continuously adapt the heat output to variations in heat demand over time. Pellet burners for use in central heating boilers were developed during the 1990s and now have a significant share of the domestic heating market. In Austria, Germany and Denmark, complete pellet boiler units are the most common appliance types. In other countries, the interest in pellet heating is just awakening.

At present, there are several types of small combustion devices for biomass pellets, which can be roughly divided into the types described below [1].

1.1 Loose stokers with integrated hopper

One possibility for converting to wood pellets is to retrofit an existing boiler with a separate pellet burner (Fig. 1). This type of pellet burning system demands a lot of space in the boiler room. When the boiler is in good condition, the size and the geometry of the combustion chamber make it suitable for combustion of wood pellets, particularly, where the existing boiler was used for burning coal. An important condition to be fulfilled to make a retrofit system work satisfactorily is

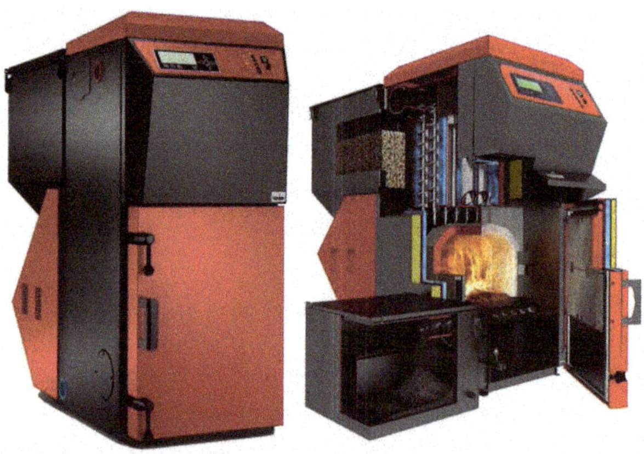

Figure 1: A pellet burner with an integrated hopper for wood pellets [2].

that the chosen pellet burner should be sized correctly so that it fits the capacity of the existing boiler.

1.2 Wood pellet boiler combined with oil-fired boiler

When installing a wood pellet boiler, it can be desirable to keep any existing natural gas fired boiler as a supplement to the wood pellet boiler in the coldest winter periods or as a spare boiler [3].

1.3 Pellet burners without hoppers

They are compact pellet burners, which resemble oil burners both in size and appearance (Fig. 2). These compact pellet burners were build upon differing principles of wood pellet firing of pellets than used in the traditional compact boiler.

This type of pellet burner has the advantage that it can be retrofitted onto an existing boiler and that the installation is simple and demands little space. However, there is a final demand that is the stoker must to have a size and shape that makes it suitable for the combustion of wood pellets [5].

1.4 Compact boiler systems

Compact boilers, also called stoker boilers, are characterized by having a boiler, a combustion system and a fuel hopper built together into one compact unit. The whole unit is designed so that the single elements fit together (Fig. 3). The hopper, combustion system and boiler are all designed to the same capacity, which is not always the case when installing a loose stoker on an existing boiler.

Some compact boilers are designed to burn wood pellets, wood chips and grain. However, shifting between different fuel types normally requires an adaptation of the operating unit to the fuel it is intended to use [6]. As with other boilers' systems

Figure 2: An example of a pellet burner that amongst others can be used for the retrofitting of an existing boiler [4].

Figure 3: Typical compact boiler system where hopper, combustion system and boiler are built together into one unit [4].

that burn solid biofuel, the compact boiler system takes up more space than an oil-fired boiler with the same power output.

1.5 Pellet stoves

There are two main types of pellet stoves available: stand-alone pellet stoves and chimney integrated stoves. The only difference between these two types is that the latter is especially dimensioned to be placed in an open fireplace. The most common stove is the stand-alone stove. Stand-alone stoves usually have an integrated pellet storage, which allows storage of a limited quantity of pellets, usually enough for 1 or 2 days. Based on the wood stove, pellet-fired stoves designed for installation in the living room have been developed and have become increasingly popular in recent years. In comparison to normal wood stoves, pellet stoves are designed for house owners who emphasize a need for simple operation and a minimal need for maintenance (Fig. 4). Most pellet stoves are equipped with a thermostat and electronic ignition and therefore they do not demand manual intervention such as stoking or the controlling of their output.

1.6 Pellet stoves with back boilers

Some manufacturers have produced pellet stoves with an integrated back boiler. They produce 8–9 kW of heat, of which the water tank takes 35%–45%. This heat is used to heat other rooms in the house via the central heating system. The remainder heats the room in which the stove is placed. In comparison to wood

Figure 4: Wood pellet stove designed for installation in the living room [7].

stoves pellet stoves, they are still relatively expensive, but in return, they have some of the advantages of pellet-fired central heating installations [8].

1.7 Wood pellet boiler with solar heating

The solar heater can either be implemented so that it only delivers domestic hot water or so that the solar panel delivers both hot water and space heating. As opposed to wood pellet boilers, a solar heater demands storage for the heat, which is produced by the solar panel during the day. The capacity of the storage tank has to correspond to the area of the solar panel and in many cases; the hot water tank will have sufficient capacity [9]. The advantage of combining a wood pellet boiler with a solar heater (Fig. 5) is that in the summer period, the solar heater can completely or partly cover the domestic hot water demand, so that the wood pellet boiler can be taken out of service [10].

2 Wood Pellet Boilers

2.1 Technical features

Pellet-fired boilers (Fig. 6) allow continuous automatic combustion of a refined and well-defined fuel, with the burning rate controlled by the rate of fuel supply rather than by restricting the primary air. The fuel is fed automatically into the

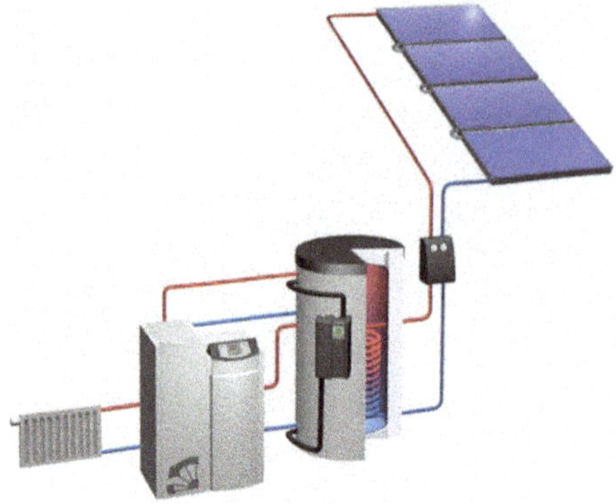

Figure 5: Combined wood pellet boiler and solar heating [11].

Figure 6: Two wood pellet boilers [12].

combustion chamber by means of an auger from a storage hopper. Some burners are equipped with a smaller pellet storage (enough for one or a few days of operation) that can be refilled manually or by an automatic system from a larger storage. The combustion air is supplied by an electric fan that may also provide a distinct secondary airflow. The fuel feed is typically by means of an auger from a storage hopper. The fuel delivery rate can normally be fine-tuned by on/off timers causing the auger to cycle over a short interval, with the percentage of on-time varied depending on the heat demand. Operation is normally controlled by a thermostat in the heated space, or by an aqua-stat in a boiler or heat storage tank. Ignition is by means of an electric device or by maintaining a pilot flame. Minimal maintenance is desired and the best burners can be in operation for more than a week without the manual removal of ash, etc. In fact, the most important development is perhaps to increase the availability and decrease the necessary maintenance. For instance, burners can be installed in boilers with automatic ash removal.

Modern wood pellet-fired systems are designed for automatic operation, so that the supply of fuel and combustion control only requires the minimum of manual intervention. Because of this, wood pellet boilers have advantages in terms of comfort compared with other solid fuel boilers. In full, a pellet-fired boiler consists of: fuel storage, fuel hopper, stoker or wood pellet burner, boiler, ash outlet and ash container, systems for controlling the supply of fuel and air, and flue gas exit and chimney.

2.1.1 Fuel storage

The design of the fuel storage system depends on the space the consumer has available for it and the wood pellet delivery method chosen.

- *Storage in bags*: Some domestic boilers use wood pellets that are delivered in bags with a weight of 10–25 kg. Even though the bags are made out of a plastic, they have to be stored dry because the pellets are able to absorb moisture from the air. If wood pellets are exposed to moisture, for example in the form of rain, they will decompose and become useless as a fuel. The bags, therefore, have to be stored in a dry place, for example in an annex or in the basement. From the storage room the bags are carried to the boiler room, where they are manually emptied into to the fuel hopper.
- *Storage in loose weight*: Larger fuel stores where the wood pellets are delivered by lorry and blown into the storage room can be arranged in a room neighbouring the boiler room, in a silo or a room that is arranged in an attic above the boiler room. In such cases, special fire protection measures for the storage room must be taken. Storage of loose wood pellets by the consumer will typically take place in a pellet silo as can be seen in Fig. 7. A pellet silo will typically contain 5–6 m^3 of pellets. This type of storage gives the least

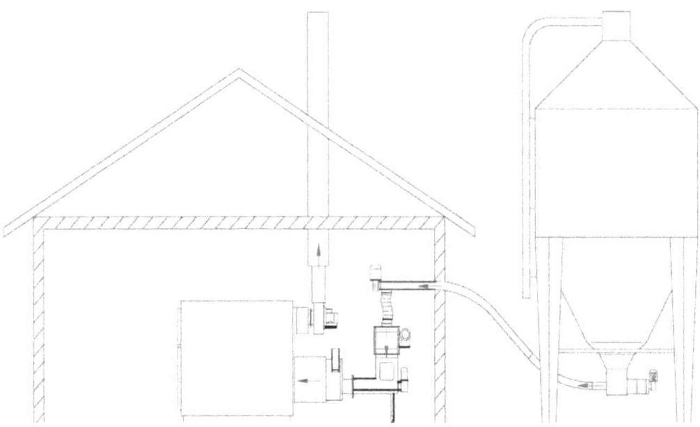

Figure 7: Drawing of a complete boiler plant with an outdoor silo for wood pellets [4].

possible amount of work and only demands an agreement with the supplier for the regular delivery of new pellets as the old ones are being consumed. The silo has to have steep side walls towards the screw conveyor. If the sides are not sufficiently steep, the pellets will roll down, but the dust will remain in the silo. This means that after a while the boiler will only be firing dust, which will increase the boiler temperature and reduce efficiency. The increased temperature also increases the risk of the ash clinkering.

Independent of the storage system selected, the room in which the pellets are stored has to be dry and the creation of dust has to be minimized.

The maximum content of moisture in the raw material at 15% is so low that dry rot and bacteria are not active. Because of this, wood pellets are in principle sufficiently dry to be stored for an indefinite amount of time without being decomposed by microorganisms.

An important objective when handling wood pellets is that the pellets should be exposed to as little physical stress as possible, because it can cause the creation of dust and small particles. During mechanical handling of wood pellets, small parts are loosened and thereby dust and small particles are created. Usually, pellet manufacturers remove dust and small particles with a screen or exhaust equipment. However, it is difficult to avoid the creation of dust during handling and storage. To reduce dust nuisance, it is important that bags and wood pellets are handled in a way that does not cause the wood pellets to break. If the fuel storage room is established indoors, it is recommended that a cyclone is mounted in the injection system to reduce the dust nuisance in the storage room. Low pressure in the fuel storage room during the injection of wood pellets furthermore ensures that the dust does not spread to other parts of the building. Dust is not only annoying and harmful to the health but can also constitute a fire risk and cause dust explosions. Thus, regular vacuum cleaning and removal of dust is important to reduce its associated problems such as: uneven combustion caused by uneven induction of fuel; low efficiency; high content of unburnt fuel in the ash and diffusion of dust.

2.1.2 Fuel transport

Transportation of wood pellets from the storage area to the fuel hopper usually takes place through a system consisting of screw conveyors. When choosing the conveyors, it is important that they are suited for the transportation of wood pellets, which mainly means that they rotate slowly. Quickly rotating conveyors act as whisks and break the wood pellets.

2.1.3 Fuel hopper

The fuel hopper acts as a store from which the wood pellets are automatically transported with a screw conveyor into the combustion chamber. The fuel hopper has to be made of non-flammable material and has to be able to be closed with an air tight lid. The fuel hopper normally holds an amount of fuel that corresponds to approximately one day's consumption at full load.

2.1.4 Stoker or wood pellet burner

From the fuel hopper, the firing of the pellets happens via a stoker or screw conveyor (or a corresponding device) that conveys the wood pellets into the combustion chamber of the boiler. The stoker is driven by a motor that is controlled so that it delivers the amount of fuel that corresponds to the heat consumption from the boiler or to the wanted boiler output. Inside the stoker pipe, a temperature sensor is placed that activates a sprinkler device if the temperature in the stoker pipe gets too high, e.g. due to back firing. The sprinkler device is supplied with water from a water tank placed on the boiler or the stoker.

Pellet burners for domestic use are usually constructed for a nominal thermal output of <25 kW. Depending on the feeding system, three different types of pellet burners can be identified (Fig. 8) with the flame burning either horizontally or upwards. Perhaps, the most interesting burners are the horizontally fed ones, in which the gases are released within a combustion chamber (tube) during devolatilization and in which a primary combustion zone can be defined, in contrast to the other types of burners. To ensure minimal emissions, it important to obtain stable combustion conditions and minimal stand-by periods, i.e. by avoiding the number of start-ups or operation with a pilot flame. In addition, the overall emissions are generally reduced when using electric ignition rather than a pilot flame.

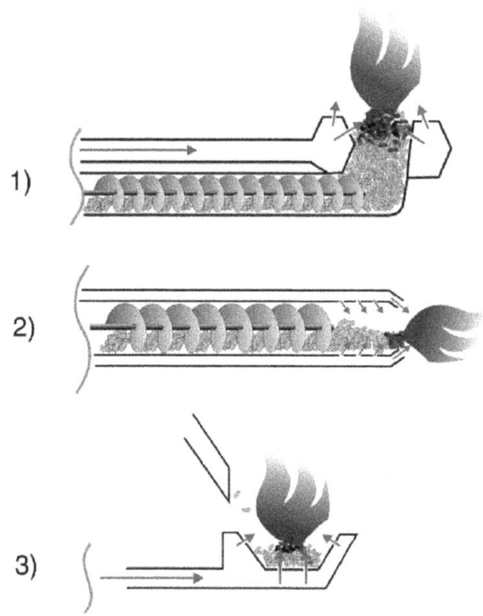

Figure 8: Wood pellet burners: (1) underfed burners; (2) horizontally fed burner and (3) overfed burner [13].

In some Nordic countries, it is quite common to replace an oil burner in an existing heating boiler with a pellet burner. In this case, it is important that the pellet burner is properly designed for the boiler. If the flame hits a cold surface, high emissions of soot and hydrocarbons and a low efficiency will result. In addition, the flue gas flowing from a pellet burner is higher than that of oil, and the installed thermal output of a pellet burner should therefore be less than that of a corresponding oil burner. This is important since the residence time in the boiler may otherwise be too short, resulting in flue gas temperatures that are too high and emissions of unburned hydrocarbons. This means that a consumer can get slightly less heat output from an oil-fired boiler converted to use a pellet burner. Normally, this does not cause any problems because, for technical reasons, oil boilers are normally over-sized for small houses. Some burners are equipped with an integrated smaller pellet storage (enough for one or a few days of operation) that can be refilled manually. Other burners may have an automatic system, connected to a larger storage. The fuel is usually fed to the burner by means of a screw, but other feeding systems also exist.

2.1.5 The boiler

The boiler consists of a combustion chamber with a water jacket around it. In the combustion chamber, the wood pellets are combusted, with air supplied from one or more fans.

The combustion chamber may be designed in various ways. In smaller boilers, it is fairly normal for the combustion to take place in a burner pipe on the stoker itself. The combustion air is supplied through a series of openings in the burner pipe. In larger boilers, combustion normally takes place on a grate.

The combustion chamber may have a lining made of a fire-resistant ceramic material or cast iron, which partly contributes to keeping a high temperature and is also more resistant to heat stress from the combustion. To ensure good combustion the air supply is very important. When combusting wood pellets, it is especially important that the air is supplied in such a way that the gasses produced during combustion get sufficient oxygen to be combusted completely and before they leave the combustion zone. Water from the radiator system of the house circulates in the water jacket around the combustion chamber. The heat that is generated during combustion is transferred to the boiler water through the heating surfaces in the boiler. A central heating pump pumps the heated boiler water to the radiators. In some cases, the heat is transferred to the radiator water via a heat exchanger. Another heat exchanger ensures the heating of domestic hot water.

2.1.6 Ash container and ash outlet

When wood pellets are combusted, the ash generated is generally low (0.5%–1%). A larger amount of ash than that may be a sign that the pellets are of poor quality. For example, they may contain an amount of impurities in the form of sand or other inorganic materials. If the ash creates burnt cakes or slag, it may either be because of impurities in the pellets or too high combustion temperature.

In pellet appliances, the ash is collected in an ash dump or in an ash container, which is emptied by hand. Large boilers have an automatic ash outlet where the ash is transported to a container via a screw conveyor. To avoid the danger of fire, the ash has to be collected in a metal container with a lid. It may never be put into a rubbish bag or the like before, it is known for a fact that no embers are left.

2.1.7 Controlling air and fuel supply

Wood pellet-fired boilers are different from oil-fired and gas-fired boilers in one significant respect: fossil fuelled boilers operate in short heating intervals (on–off operation), as the supply of fuel and air is turned on and off momentarily. In oil-fired and gas-fired boilers, the combustion in the combustion chamber of the boiler stops at the moment that the supply of fuel is discontinued.

In wood pellet-fired boilers, on the other hand, the fuel is supplied in a steady flow. As long as fuel and air is supplied, the combustion of wood pellets will continue. If the supply is cut off, the combustion will continue until all the fuel in the combustion chamber has been converted. Heating with wood pellets thus gives longer start-up and shut-down times than is usual in oil or gas firing.

The purpose of the boiler control system is to control the fuel and air supply, so that the boiler delivers the desired heat to the heat distribution system, i.e. the radiators and supply of domestic hot water. The more heat that is needed, the more fuel has to be supplied and the more air has to be added to ensure clean combustion.

The control unit may be built upon a thermostatic control unit that registers the temperature of the boiler water at a suitable place in the system. When the temperature decreases beneath a certain level, a start-up impulse is sent to the motor that conveys the pellets into the boiler. At the same time, the control unit makes sure that the air supply corresponds to the amount of fuel that has to be combusted. Newer boilers are often fitted with an oxygen control unit based on a so-called lambda probe. This is a sensor that is placed in the flue gas duct and registers the oxygen content in the flue gas. The controlling system uses the signal from the lambda probe to ensure that a suitable amount of air is supplied into the combustion zone. The control unit is very often equipped with pre-settings that are activated when the boiler starts up and at different output levels (modulating control), for example, it can be pre-set for low, medium and high output, respectively. The control unit is of high importance. Precise control is important to obtain the optimal fuel economy and clean combustion. Finally, control is important for the sake of the easy operation of the boiler. Thus, the improvement of boiler control systems is an on-going process.

2.1.8 Flue gas outlet and chimneys

To obtain an efficient combustion, it is important that the flue gas outlet and chimney are dimensioned to ensure the ‘breathing’ of the boiler. If the chimney is blocked or wrongly dimensioned, it may lead not only to inefficient combustion but also in the worst-case scenario to the seepage of carbon monoxide [14].

The boiler must always be provided with flue gas ducts and chimney that comply with the requirements that the boiler manufacturer has stated and those of the relevant building regulations. These relate to both the diameter of the chimney pipe and the chimney height.

When the wood pellet boiler is being installed as a replacement for a previous oil-fired or gas-fired boiler, or when a loose pellet burner is installed to an existing boiler, attention must be paid to the fact that a wood pellet boiler demands a chimney diameter of no <15 cm, corresponding to a cross section area of 175 cm^2. To obtain the right exhaust conditions, it might be necessary to put up a new chimney, for example a steel chimney, or to rebuild the existing chimney for example by insertion of an insulated exhaust pipe.

Both the flue gas outlet from the boiler to the chimney and the chimney itself has to be kept clear of carbon depositions.

Problems can arise when house owners shift to a very efficient wood pellet boiler. Chimneys that have worked satisfactory with old boilers are corroded by tarry soot when a more efficient boiler is connected. New, effective wood pellet boilers have a low smoke temperature. Thus, these boilers set demands just as high on the chimney in relation to insulation and diameter as an efficient oil-fired boiler. Tarry soot in the chimney is destructive and foul-smelling and may cause a chimney fire if not stopped in time. Tarry soot is a result of condensation that develops, when the water vapour in the flue gas is cooled down below approximately 50°C. Condensation happens in the coldest places of the chimney – typically at the top of the chimney or in the part that passes a cold, poorly insulated attic. The condensation is absorbed by the joints and bricks of the chimney and dissolves the carbon it meets. The dissolved carbon then migrates through the walls of the chimney. After some time, brown and black–brown discolorations will be visible on the outside of the chimney and in the attic at the horizontal divisions.

The reason for the carbon layer is typically a combination of a poorly insulated chimney, low flue gas temperature and incomplete combustion. When closing the air supply, the fuel lasts longer, but in return unburned flue gasses will settle in the plant and in the chimney as carbon deposits. When the air supply is shut off, the flue gas speed will decrease and the water vapours in the flue gas will stay in the chimney and mix with the unburned flue gasses. When this has happened, it is often seen in the chimney as a layer of carbon with a glossy surface (shining soot).

To prevent condensation, the chimney can be insulated on the inside by lowering an insulating interior into the chimney or casting an insulating interior inside the old chimney. It is also possible to lower a steel-lining into the chimney. Installing a chimney liner results in a chimney with a smaller diameter [15].

2.2 Safety requirements regarding back-burning

Preventing burn-back into the feed auger and storage hopper is a major safety concern for pellet-fired equipment. Burners and stoves can (and should) be equipped with a number of independent safety systems such as:

- A sprinkler system.
- A drop chute into the combustion chamber.
- A thermal sensor in the feed system.

- An airtight fuel storage hopper.
- A double auger feed system with a firebreak between the two augers.

2.3 Emissions

Emissions from residential wood burning appliances are in general influenced by combustion technology and process conditions and by fuel properties. Current appliances with low emission levels tend to have a clear delineation of primary and secondary air supply to maintain an air-starved condition in the fuel bed for output control, but still permit adequate burnout of combustible gases. Introducing secondary air under conditions of adequate temperature for ignition and burnout can markedly reduce emission levels, and various strategies have evolved to achieve this objective [16].

Good pellet burners show very low emissions levels of hydrocarbons and carbon monoxide. However, NO_x emissions are significant despite the rather low nitrogen content in the pellets. In fact, the conversion of fuel nitrogen to NO_x is in many cases close to 100% (Fig. 9).

The amount of excess air in the secondary zone is not only of importance for carbon monoxide (CO) and unburnt hydrocarbons (OGC), there is also a trade-off between these emissions and the emission of nitrogen oxides (NO_x): too little air will result in increased emissions of CO and OGC, but will keep the amount of NO_x in the flue gas small; with greater excess air, more NO_x will be released from the burner. Since the emissions of NO_x usually increase as the emissions of unburned hydrocarbons decrease (Fig. 10), the challenge is to balance efficient NO_x reduction with low emissions of hydrocarbons. In addition, particulate emissions from small-scale pellet combustion can be significant.

2.4 Efficiency

The efficiency of pellet combustion devices is typically within the range 60%–90%, but the final value is largely dependent on its design and on the fuel properties.

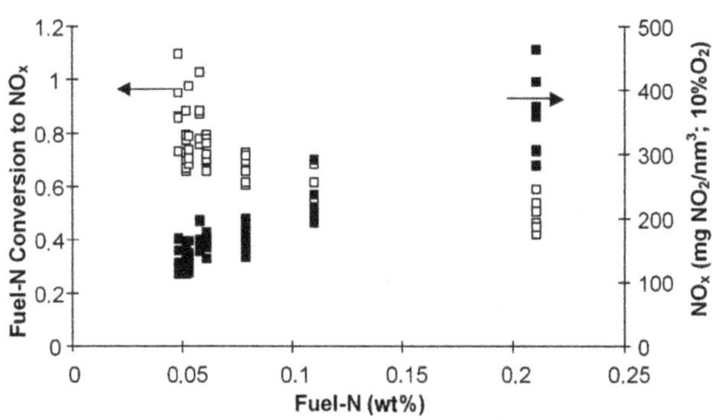

Figure 9: The conversion of fuel nitrogen to NO_x in wood pellet combustion [17].

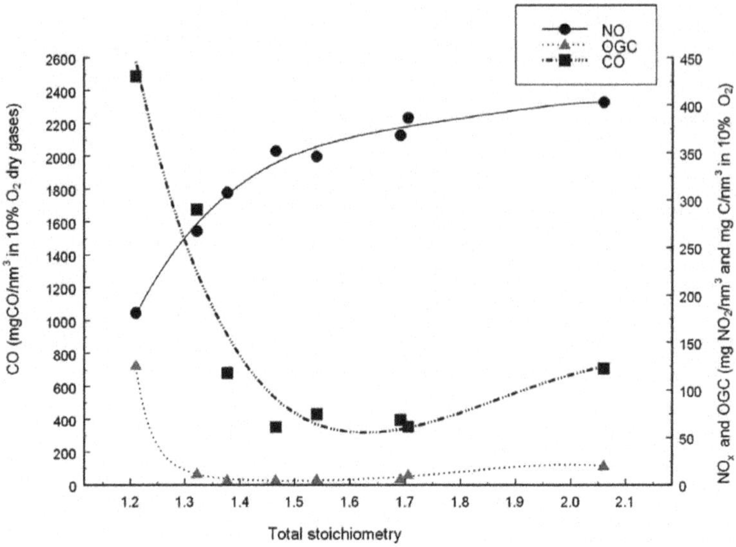

Figure 10: Correlation between unburned compounds (CO and OGC) and NO_x. OGC are all gaseous hydrocarbons that can be measured using a flame ionization detector (FID) analyser at 200°C [13].

Three types of losses are inherent: (1) latent heat losses from the moisture in the flue gas, (2) losses due to unburnt fuel and (3) sensible heat loss in the flue gas.

Maximizing efficiency has traditionally been dealt with by providing a large heat exchange surface area relative to heat output. While this results in adequate efficiency for many applications, it does not address high levels of unburned hydrocarbons and carbon monoxide, which reduce efficiency and result in high pollutant emissions relative to conventional heating systems.

A major impediment to the design of good appliances has been the lack of commonly accepted test protocols for determining the emissions and efficiency of these appliances. It is only since the 1990s that consensus standards for these evaluations have been developed, and they continue to undergo refinement as more experience with their use is gained [18].

2.5 Certification

A burner must fulfil a number of technical requirements as well as meet tough emission standards [19]. In North America, the manufacturers have developed a safety certification standard for pellet stoves. Certification bodies typically use this standard together with additional in-house requirements. Underwriters' Laboratories of Canada is in the process of developing a consensus standard for pellet stoves for that country. Requirements for residential central heating systems and larger pellet equipment are covered in the CSA (Canadian Standards Association) standards that also cover firewood-fuelled systems, and do not address efficiency or emissions. The US exempts most pellet stoves from emission requirements due to their high excess air level.

3 Wood Pellet Stoves

Special kinds of stoves have been constructed to combust only pelletized material. Contrary to conventional wood stoves, pellet stoves are dependent on electricity for their operation. An electric fan controls the combustion process by varying the supply of combustion air. This results in low CO and C_xH_y emissions. Particulate emissions are generally low and consist mainly of inorganic material in contrast to wood stoves where particle emissions are dominated by soot and tar. Pellet stoves are typically provided with small fuel storage, fuel feeder, combustion air blower, burner shell and a flue gas discharge system. They are equipped with a rather extensive control system. Figure 11 shows the principle of a pellet stove with a maximum nominal power output of 10 kW and a turn-down ratio of a factor of 4 (2.5–10 kW).

Pellet stoves work using the same basic principles as pellet boilers. The pellets are combusted in an integrated burner, which is similar to the ones used in pellet boilers. Most pellet stoves use a fall channel from the integrated or external storage to feed the pellets to the burner pot. Through openings in the bottom of

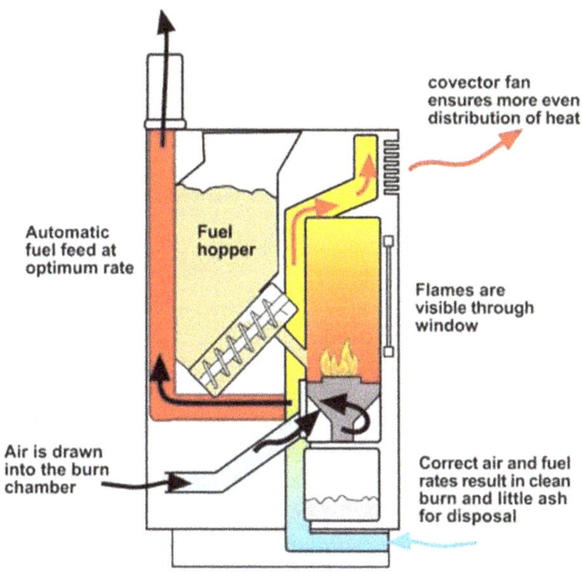

Figure 11: Diagram of a typical stove [20].

the pot the primary air and the hot air for the automatic lighting is supplied [21]. The secondary air is usually preheated through the mantel of the pot and fed by many small openings of the mantle. The aspirator supplying the combustion air for the stove is placed below the burner. Sometimes an additional fan is used to improve the heat transfer from the stove to the ambient air. To simplify ash removal, the pellets are combusted on a manually or automatically operated moveable grate plate allowing the ashes to fall down in the ash container [22].

The fuel hopper is filled from the top. The burning time with a full hopper at nominal power output is in the order of 6 h. For transport of the fuel to the burner, a fuel-feeding system is used. The speed of the feeder is related to the heat demand. The burner of the stove consists of a shell-like pot in which the pellets are burned. Combustion air is drawn through slots in the inner wall of the shell by means of an exhaust fan, which reduces the pressure in the combustion chamber. The flue gases are discharged at the top of the combustion chamber and in this example flow down towards the chimney constructed in the bottom of the stove, passing through a heat exchanger on the way. Room air enters the stove near the bottom of the unit and leaves the stove at the top.

References

[1] Obernberger, I. & Thek, G., *The Pellet Handbook: The Production and Thermal Utilisation of Biomass Pellets*, Routledge, 2010, p. 549.

[2] http://ashburnstoves.co.uk/index.php/ourproducts/automatic-boilers/, 2014.

[3] Rabaçal, M., Fernandes, U. & Costa, M., Combustion and emission characteristics of a domestic boiler fired with pellets of pine, industrial wood wastes and peach stones. *Renewable Energy*, **51**, pp. 220–226, 2013.

[4] Pelletsatlas, *English Handbook for Wood Pellet Combustion*, European Biomass Industry Association, 2009.

[5] Eskilsson, D., Rönnbäck, M., Samuelsson, J. & Tullin, C., Optimisation of efficiency and emissions in pellet burners. *Biomass and Bioenergy*, **27(6)**, pp. 541–546, 2004.

[6] Demirbaş, A., Biomass co-firing for coal-fired boilers. *Energy, Exploration & Exploitation*, **21**, pp. 269–278, 2009.

[7] http://www.archiexpo.com/prod/consolar/contemporary-pellet-wood-stoves-62603-172615.html, 2014.

[8] http://www.energysavingtrust.org.uk/Generating-energy/Choosing-a-renewable-technology/Wood-fuelled-heating/Wood-fuelled-stoves-with-back-boilers

[9] Persson, T., Nordlander, S. & Rönnelid, M., Electrical savings by use of wood pellet stoves and solar heating systems in electrically heated single-family houses. *Energy and Buildings*, **37**, pp. 920–929, 2005.

[10] Fiedler, F., Nordlander, S., Persson, T. & Bales, C., Thermal performance of combined solar and pellet heating systems. *Renewable Energy*, **31**, pp. 73–88, 2006.

[11] http://www.celsico.com/products/celsico-wood-pellets/combustion/pellets-combined-with-solar-energy.html, 2011.

[12] http://www.frolingenergy.com/services/wood-pellet-boilers/, 2013.

[13] Van Loo, S. & Koppejan, J., *Biomass Combustion and Co-Firing*, Earthscan: London and Sterling (VA), 2004.

[14] Bordzilowski, J. & Darowicki, K., Anti-corrosion protection of chimneys and flue gas ducts. *Anti-Corrosion Methods and Materials*, **45**, pp. 388–396, 1998.

[15] Öhman, M., Nordin, A., Hedman, H. & Jirjis, R., Reasons for slagging during stemwood pellet combustion and some measures for prevention. *Biomass and Bioenergy*, **27**, pp. 597–605, 2004.

[16] Peterschmidt, E. & Taylor, M., Boilers and boiler control systems. In *Encyclopedia of Energy Engineering and Technology*, Taylor & Francis, pp. 93–103, 2007.

[17] Eskilsson, C., Rönnbäck, D., Samuelsson, M. & Tullin, J., Optimisation of efficiency and emissions in pellet burners. *Biomass and Bioenergy*, **27**, pp. 541–546, 2004.

[18] Lévesque, B., Allaire, S., Gauvin, D., Koutrakis, P., Gingras, S., Rhainds, M., Prud'Homme, H. & Duchesne, J.F., Wood-burning appliances and indoor air quality. *Science of the Total Environment*, **281**, pp. 47–62, 2001.

[19] P-marking pellet burners, Certification Rules CR 029, SP Technical Research Institute of Sweden, www.sp.se

[20] http://www.bioenergyinc.ca/WoodFuel/, 2014.

[21] Win, K.M., Persson, T. & Bales, C., Particles and gaseous emissions from realistic operation of residential wood pellet heating systems. *Atmospheric Environment*, **59**, pp. 320–327, 2012.

[22] Sopha, B.M., Klöckner, C.A., Skjevrak, G. & Hertwich, E.G., Norwegian households' perception of wood pellet stove compared to air-to-air heat pump and electric heating. *Energy Policy*, **38**, pp. 3744–3754, 2010.

CHAPTER 6

Gaseous Emissions from the Combustion of Biomass Pellets

Jose Antonio Perez-Jimenez
Department of Biomass, CTAER Andalucia Foundation, Scientific and Technology Park, Jaen, Spain.

Abstract

Biomass is a sustainable energy source with significant potentials for replacing fossil fuels and electricity for heating purposes. Present residential wood combustion can, however, be a significant source of ambient urban air pollutants such as volatile organic compounds, polycyclic aromatic hydrocarbons and particulate matter. Ambient exposure to these pollutants in general has been associated with different health effects such as cardiopulmonary disease/mortality and cancer risks [1–3]. Wood pellets are generally a clean, dry and easily fed fuel to be used in special boilers, burners and stoves on the residential market. Compared with other modern technologies, the majority of the wood-fired appliances currently used suffer from poorly optimized conditions, resulting in considerable emissions of products from incomplete combustion. New and upgraded biomass fuels provide possibilities of more controlled and optimized combustion with less emission of PIC [4]. For future health impact assessments, regulatory standards and evaluations concerning present and future residential biomass combustion, as well as a solid qualitative and quantitative knowledge of the emissions from different sources, are of vital importance. In consequence, there is still a strong need for detailed characterization and quantification of the emissions from residential biomass appliances when using different fuels and combustion techniques [5].
Keywords: Pellets, biomass, emissions, pollutants, combustion.

1 Introduction

The environmental impacts of air pollution from most biomass combustion appliances today are far from negligible. However, they show significant advantages in comparison to those that use fossil fuels [6].

Table 1: Average energy consumption at procurement of different fuels [8].

	Energy consumption at procurement	
	MJ/tonne	% of energy content
Native wood pellets	627	3.6
Imported wood pellets	787	4.5
Natural gas	2,840	5.8
Oil	4,617	11.4
Coal	1,764	6.7

Biomass is a renewable fuel considered to be CO_2-neutral with respect to the greenhouse gas balance. However, this is only true if we are able to achieve very low levels of emissions from incomplete combustion [7], and if we do not consider the energy consumption from fossil fuels in biomass treatment and logistic processes such as harvesting, transportation or compaction (Table 1).

To evaluate the real environmental impacts of biomass combustion, a life cycle assessment (LCA) should be carried out. This type of evaluation includes the various stages of the life cycle of the biomass, from procurement of the fuel, transportation, storage and conversion to the discharge and handling of ashes [9]. The construction, operation, maintenance and decommissioning of the energy converting technology should also be included in the assessment according to the LCA method. The exchanges to the surrounding environment in terms of emissions to air, soil and water are then inventoried for each stage in the life cycle. These exchanges are added up to indicate environmental impacts such as global warming, acidification and ozone depletion in the environmental impact assessment. Therefore, it is not fully correct to define biomass fuels as CO_2-neutral when considering the complete life cycle [10].

Emission reduction measures are known and available for all harmful emission components, and whether the emission reduction measures are implemented or not merely depends on emission limits and cost-effectiveness. Technology to reduce emissions and improve combustion efficiency is being developed continuously. Wood burning appliances have gradually been improved because of the permanent R&D activities that have been developed during the last years and that has led to an advanced understanding of the combustion process as well as of emission formation mechanisms. Better knowledge about the basic mechanisms governing the process also has led to a more focused further development of primary measures for emission reduction.

2 Gaseous Emission Components from Biomass Combustion

Gaseous emissions from biomass combustion applications can be divided into two major groups: emissions from complete combustion and emissions from incomplete combustion.

Table 2: Average proximate analysis of wood pellets and combustion products [11].

Matter	Share	Combustion product
H_2 – Hydrogen	5.8	H_2O – Water vapour in flue gas
C – Carbon	46.5	CO_2 – Carbon dioxide/carbonic acid gas
O_2 – Oxygen	39.5	O_2 – As surplus oxygen
A – Ash	0.9	Ash – Partly as dust in the flue gas
S – Sulphur	0.05	SO_2 – Sulphur dioxide
N_2 – Nitrogen	0.28	N_2 – Nitrogen in the flue gas
H_2O – Water	7.0	H_2O – Water vapour in the flue gas

2.1 Gaseous emissions from complete biomass combustion

The amount and composition of flue gas created during the complete combustion are determined by the proximate analysis of the wood pellets. Table 2 shows the typical result of a proximate analysis of wood pellets, together with the combustion products that each batch creates during complete combustion. Nevertheless, the concentration of gases in the flue gas will depend on the fuel and for some components also on the design of the combustion system.

2.1.1 Carbon dioxide

Carbon dioxide (CO_2) plays a significant role in the greenhouse effect and thus the emission of CO_2 to the atmosphere is problematic. Since wood pellets and other biofuels contain carbon, naturally CO_2 is also created during their combustion. However, when combusting wood fuels there is no more carbon dioxide released than that bound up in the wood as carbon and the only CO_2 attributable to pellets is that emitted during their production and transport. This results in the almost zero CO_2 emissions from biomass fuels shown in Table 3 in comparison with fossil fuels.

Carbon constitutes approximately 50% of completely dry wood, which corresponds to 47% in wood pellets with a moisture content of 7% (Table 2).

2.1.2 Nitrogen oxides (NO_x)

NO_x is a generic term for the sum of the gases NO and NO_2. NO_x is a greenhouse gas that at the same time contributes to the acidification of precipitation. Nitrogen oxides are created partly during the combustion of fuels with a natural nitrogen content including biofuels and partly in the boiler room by the nitrogen that is injected during combustion with the combustion air. The design and size of the boiler combustion chamber is therefore of importance in the emission of nitrogen oxides. It is not possible to give completely plain guidelines for the design of boilers with low NO_x emission but the following tendencies are certain: high N-content in the fuel increases NO_x emission and high combustion temperatures generate high NO_x emissions [13].

Table 3: Gaseous emission key figures for delivered energy [12].

Fuel	CO_2 (g/MJ)	NO_x (mg/MJ)	SO_2 (g/MJ)	CO (mg/MJ)	TOC (mg/MJ)
Wood pellets	0.108	130–300	0	50–3000	<10
Straw	0.108	130–300	13	500–3000	<10
Oil	0.972	75	94	15–30	0–2
Natural gas	0.72	50–100	0.5	15–20	0–2

NO_x emissions from biomass combustion applications are mainly a result of complete oxidation of fuel nitrogen, in contrast to fossil fuel combustion applications where nitrogen in the air also contributes, to some extent, to the NO_x emission level. The possible gas-phase reaction mechanisms for NO_x formation in biomass combustion applications are as follows:

- *The fuel NO_x mechanism [14–16]*: Fuel nitrogen is converted to NO (>90%) and NO_2 (<10%) through a series of elementary reaction steps called the fuel NO_x mechanism. Important primary nitrogen-containing components are NH_3 and HCN. However, significant amounts of NO and N_2 may also be found in the pyrolysis gas. If sufficient O_2 is available, NH_3 and HCN will mainly be converted to NO through different reaction routes. However, in fuel-rich conditions NO will react with NH_3 and HCN, forming N_2. This is utilized as a primary NO_x reduction measure. By optimizing the primary excess air ratio, temperature and residence time, a maximum conversion of NH_3 and HCN to N_2 can be achieved.
- *The thermal NO_x mechanism [17,18]*: Nitrogen in the air starts to react with O radicals and forms NO at temperatures above approximately 1300°C. The amount of NO formed increases with increasing temperature, O_2 concentration and residence time. However, in biomass combustion applications, the combustion temperatures are, in general, lower than 1300°C. Thermal NO_x is a post-flame problem, *meaning that the major formation of thermal NO_x occurs in the post-*flame gases, after the main combustion, due to its dependence on O_2 concentration and residence time.
- *The prompt NO_x mechanism [14–19]*: Nitrogen in the air may also react with CH mainly forming HCN, which then follows the reaction steps of the fuel NO_x mechanism. The prompt NO_x mechanism is less temperature-dependent and much faster than the thermal NO_x mechanism. However, it is only important in fuel-rich conditions and is very dependent on the CH concentration. The prompt NO_x mechanism has not been found to be of significant importance in biomass combustion applications, in contrast to fossil fuel combustion applications.

Additionally, fuel nitrogen is retained in the char and is largely oxidized to NO in the char combustion phase but may subsequently be reduced to N_2 by a fast

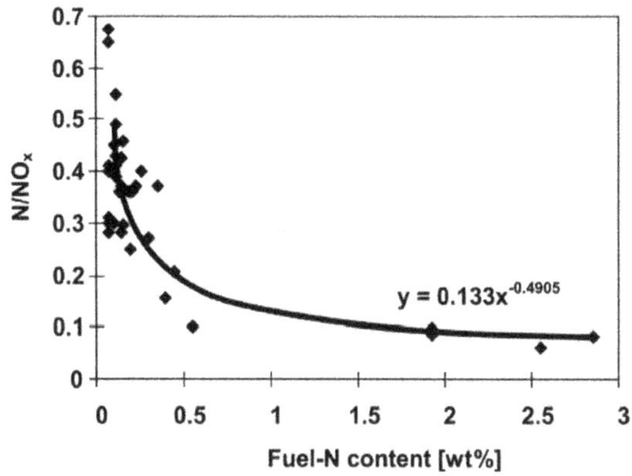

Figure 1: Fraction of fuel nitrogen converted to NO_x for various wood fuels in various wood combustion applications, as a function of fuel nitrogen content, together with a trend line [22].

heterogeneous reaction with the char [20]. The amount of fuel nitrogen retained in the char relative to the amount of fuel nitrogen released in the devolatilization phase is determined in part by the thermal exposure of the fuel [21]. The fuel NO_x emissions increase with increasing nitrogen content in the fuel, excess air ratio and combustion temperature, up to a point where all fuel nitrogen intermediates have been converted to either NO_x, N_2O or N_2. However, the fraction of fuel nitrogen converted to NO_x decreases with increasing nitrogen content in the fuel (Fig. 1). Figure 2 illustrates the relative importance of the fuel, thermal and prompt NO_x mechanisms as a function of combustion temperature. NO_x emissions may be reduced by both primary and secondary emission reduction measures.

2.1.3 Nitrous oxide

Nitrous oxide (N_2O) emissions are a result of complete oxidation of fuel nitrogen. The N_2O emission levels from biomass combustion are very low but they do contribute to some degree to the greenhouse gas effect because of the high global warming potential factor of N_2O [24].

2.1.4 Sulphur oxides

Sulphur oxides (SO_x) are a result of complete oxidation of fuel sulphur. It is mainly SO_2 (>95%), but some SO_3 (<5%) may be formed at lower temperatures. The emission of SO_2 can be significantly reduced by means of pellet firing in comparison with those emissions produced from fossil fuels systems (Table 3).

In has to be highlighted that the sulphur in the fuel will not be completely converted to SO_x: a significant fraction will remain in the ashes while a minor fraction is emitted as a salt (K_2SO_4) or as H_2S at lower temperatures. Measurements at two district heating plants in Denmark using straw as a fuel showed that 57%–65% of

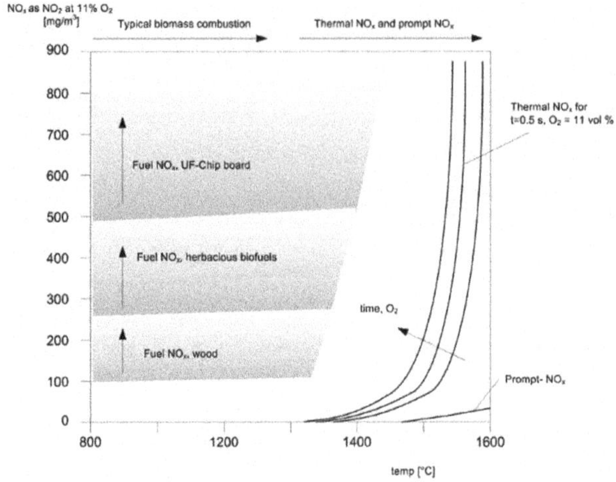

Figure 2: Fuel-NO_x emission levels as a function of temperature and fuel type (fuel-N content) and comparison with thermal and prompt NO_x formation [23].

the sulphur was released into the flue gas, while the remainder was bound in the ashes [25]. Houmøller and Evald [26] reported nine closed sulphur balances from several full-scale measurements in Denmark, four straw-fired units, three wood-chip-fired units and two pellet-fired units and they found similar results.

Special consideration needs to be given to wood pellets made with additives as binding agents. The lignin-based binders that are sometimes used in wood pellets manufacturing are mostly either Wafolin or lignosulphonate, which contain approximately 6 mass percentages of sulphur. The amount of binder is usually <1% of the finished wood pellet. With a binder content of 1%, the content of sulphur in the pure wood pellet is quadrupled to approximately 0.08% compared with the content of sulphur in the pure wooden raw material at approximately 0.02%. This corresponds to a content of sulphur of 45 mg/MJ fuel, which should be compared with the sulphur content in heating oil of <25 mg/MJ [24].

2.1.5 Hydrogen chloride

Significant amounts of hydrogen chloride (HCl) may be formed from biomass fuels containing higher amounts of chlorine, such as miscanthus, grass and straw [27]. Part of the chlorine content in the fuel will be released as HCl. The fuel chlorine will not be completely converted to HCl; the main fraction is retained in salts (KCl, NaCl) by reaction with K and Na, while traces are emitted as dioxins and organic chlorine components. HCl emissions may be reduced by washing of the fuel, which is utilized to some extent for straw due to its high chlorine content [28].

2.2 Gaseous emissions from incomplete biomass combustion

Emissions caused by incomplete combustion are mainly a result of either:

- Too short residence times.
- Too low radical concentrations, in special cases, for example in the final stage of the combustion process (the char combustion phase) in a batch combustion process.
- An overall lack of available oxygen.
- Inadequate mixing of combustion air and fuel in the combustion chamber, which produces local fuel-rich combustion zones.
- Too low combustion temperatures.

These variables are all linked together through the reaction rate expressions for the elementary combustion reactions. However, in cases in which sufficient oxygen is available, temperature is the most important variable due to its exponential influence on the reaction rates. An optimization of these variables will, in general, contribute to reduce the levels of all emissions from incomplete combustion described below:

2.2.1 Carbon monoxide (CO)

Conversion of fuel carbon to CO_2 takes places through several elementary steps, and through several different reaction paths. CO is the most important final intermediate. It is oxidized to CO_2 if oxygen is available, so CO can be regarded as a good indicator of the combustion quality. CO emission is unwanted because of being an indicator of an incomplete combustion process, being toxic and combustible (a health and safety risk), and finally because CO destroys the ozone layer.

In addition, a high CO emission is often an indicator of the presence of other unwanted and dangerous matters in the flue gas. In practices, it is not possible to entirely avoid the creation of a little carbon monoxide during combustion, but its minimization is of vital importance.

Due to the fact that wood pellets are homogeneous and have very low moisture content, very low CO emissions can be achieved if the combustion conditions and the design of the boiler are the adequate ones, but huge differences can be found when the CO emissions rage are observed (Table 3). Large-scale biomass combustion systems usually have better opportunities for the optimization of the combustion process than small-scale biomass combustion applications [29]. Figure 3 shows the CO emission level as a function of combustion temperature, whereas Fig. 4 shows the CO emission level as a function of excess air ratio for various biomass combustion applications. For a given system, higher excess air ratios will result in a decreased combustion temperature while lower excess air ratios will result in inadequate mixing conditions. In addition, sufficient residence time is important to achieve low CO emission levels, mainly because CO is generally a later intermediate than hydrocarbons.

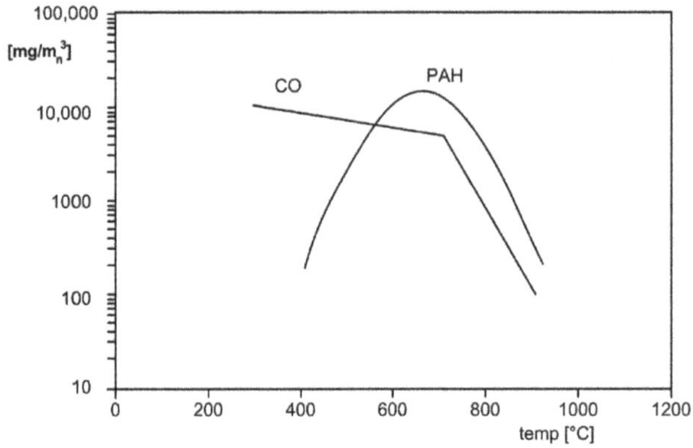

Figure 3: CO and PAH emissions in mg/m$_n^3$ as a function of combustion temperature [30].

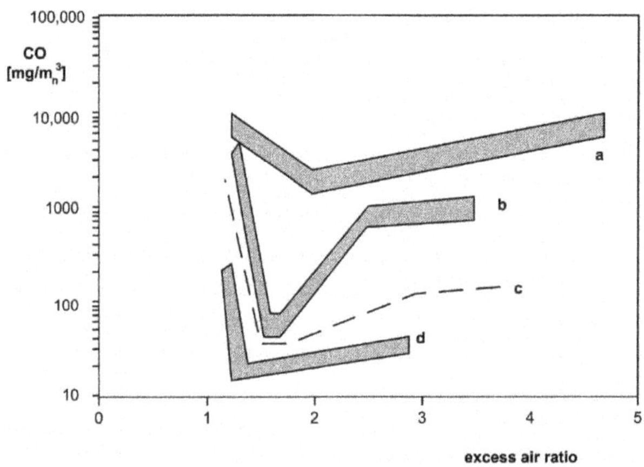

Figure 4: CO emissions in mg/m$_n^3$ as a function of excess air ratio λ [23]. (a) A simple, manually charged wood boiler. (b) A down-draught log boiler. (c) An automatic furnace with combustion technology as of 1990. (d) Automatic furnaces with appropriate combustion process control can be operated under optimum conditions.

2.2.2 Total organic compounds

Total organic compounds (TOCs) are the unburned hydrocarbon combinations in the flue gas, and their emission is undesirable because many of the organic combinations are toxic and some of them are carcinogenic [29]. Though TOC emissions are very low when biomass is combusted (Table 3), the requisites for low TOC emission are the same as for CO emissions.

2.2.3 Methane

Methane (CH_4) is usually mentioned separately from the other hydrocarbons since it is a direct greenhouse gas. In biomass combustion applications, it is an important intermediate in the conversion of fuel carbon to CO_2 and fuel hydrogen to H_2O. As for CO, emissions of CH_4 are a result of too low combustion temperatures, too short residence times or lack of available oxygen. Hydrocarbons are, in general, earlier intermediates than CO, which means that they have lower emission levels.

2.2.4 Polycyclic aromatic hydrocarbons

Polycyclic aromatic hydrocarbons (PAHs) are usually mentioned separately from other hydrocarbons due to their carcinogenic effects. They are all intermediates in the conversion of fuel carbon to CO_2 and fuel hydrogen to H_2O. As for CO emissions, PAHs are a result of too low combustion temperatures, too short residence times or lack of available oxygen. A comparison of the combustion temperature influence on the PAH and CO emission level is shown in Fig. 3.

2.2.5 Non-methane volatile organic components

This group includes all hydrocarbons except CH_4, PAH and other heavy hydrocarbons, which condense and form particle emissions. They are all intermediates in the conversion of fuel carbon to CO_2 and fuel hydrogen to H_2O. As for CO, emissions of Non-methane volatile organic components (NMVOC) are a result of too low combustion temperatures, too short residence times or lack of available oxygen.

2.2.6 Polychlorinated dioxins and furans

Polychlorinated dioxins and furans (PCDD/PCDF or PCDD/F) are a group of highly toxic components. They are found to be a consequence of the *de novo synthesis* in the temperature range within 180°C and 500°C [23]. Carbon, chlorine, catalysts (Cu) and oxygen are necessary for the formation of PCDD/F, so PCDD/F can be formed in very small amounts from all biomass fuels containing chlorine. The emissions of PCDD/F are highly dependent on the conditions under which combustion and flue gas cooling take place; therefore, wide variations are found in practice. Although herbaceous biomass fuels have high chlorine contents, their PCDD/F emissions are usually very low. This may be explained by their high alkali content, which leads to the formation of salts (KCl, NaCl) and thus to a lower level of gaseous chlorine for the *de novo synthesis*. Because of the many factors influencing PCDD/F formation, wide variations may appear even within the same biomass combustion installation, but in general, the PCDD/F emission level from biomass combustion applications using virgin wood as a fuel is well below the health risk limit.

2.2.7 Ammonia

Small amounts of ammonia (NH_3) may be emitted because of its incomplete conversion to oxidized nitrogen-containing components. This occurs in special cases in which the combustion temperature is very low. Additionally, secondary NO_x

reduction measures utilizing NH_3 injection may contribute to the NH_3 emission level due to NH_3 slippage. NH_3 emissions can be reduced by general primary emission reduction measures for emissions from incomplete combustion, and by optimizing the NH_3 injection process.

2.2.8 Ozone

Ozone (O_3) is a secondary combustion product formed from photochemical atmospheric reactions including CO, CH_4, NMVOC and NO_x. It is a direct greenhouse gas and also influences the local and regional environment. As such, it is a highly unwanted by-product of biomass combustion applications. O_3 emissions can be reduced indirectly by reducing emissions from incomplete combustion, and by NO_x emission reduction measures.

3 Emissions Thresholds in the European Union

The European legislation regarding the control of emissions has been derived from the guidelines for Large Combustion Plants (LCP) (2001/80/EG, PB L 309) as well as from the Waste Incineration Directive (WID) (2000/76/EG, PB L 332). Guideline 2001/80/EG applies for biomass, whereas guideline 2000/76/EG applies to the incineration of all other types of waste. The emission guidelines are listed in Table 4.

Table 4: Overview of LCP and WID (data in mg/m^3) [31,32].

	LCP (clean biomass, at least 50MW$_{th}$)		Waste incineration directive	
	Co-firing (6% O_2)	Stand-alone (6% O_2)	Co-firing (6% O_2)	Stand-alone (11% O_2)
NO_x (measured as NO_2)	200*	200–400	200*	200
SO_2	200*	200	50*	50
Dust	30*	30–50	10*	10
Cd + Tl			0.05	0.05
Hg			0.05	0.05
Sum heavy metals			0.5	0.5
HCl			10*	10
HF			1*	1
Dioxins/ furans (ng/m^3)			0.1	0.1
VOC			10*	10
CO			50*	50

*Mixing rule applies for composition of total flue gas.

It has to be mentioned the existence of an European standard EN-303-5 [33] and several quality labels for residential biomass heating devices that also encourage the boilers manufacturers to optimize their products and to minimize the pollutant emissions. In addition, it is important not to forget that the performance of pellet boilers in real-life conditions [34] has to satisfy the emissions and efficiency requirements of existing standards and quality labels, and that in consequence, the standard methods developed in a laboratory have to reproduce the real performance of biomass boilers to be reliable.

4 Emissions Data

The amount of pollutants emitted to the atmosphere from biomass combustion is highly dependent on the combustion technology implemented, the fuel properties, the combustion process conditions and the emission reduction measures implemented. When available, such data generally refer to a single fuel–technology combination. To obtain an objective view of emission levels from various biomass combustion applications, it is necessary to collect emission data from a wide range of fuel–technology combinations. However, due to the many parameters involved, the data can still only give an indication of typical emission levels [35].

4.1 Domestic furnaces

The results from the analysis of gaseous emissions generated in domestic appliances have shown that the traditional stove had significantly higher levels of emissions from incomplete combustion than the staged-air wood stoves and the catalytic stoves [36]. The emissions usually decrease as the size of the combustion installation increases, due to improved process control possibilities and efficient flue gas cleaning facilities. However, wide variations in the emission levels from the same stove have been reported, depending on operating conditions, characteristics of the fuel and even on the test method (forced or natural draught) [35,37,38]. In many cases, the levels of emissions from incomplete combustion may be many times higher than that reported at nominal load, usually increasing exponentially with decreasing load relative to nominal load [39]. In 1994, the Technical University of Munich performed an extensive measuring programme on emission levels from domestic wood applications [40]: wood stoves, fireplace inserts, heat-storing stoves, pellet stoves and catalytic wood stoves.

4.2 Industrial appliances

Several authors [41,42] have compiled and presented typical ranges of emission levels from various automatic wood furnaces (under stoker furnaces, grate firings and dust firings) and different biomass fuels. As in the case of domestic furnaces, the differences among the data obtained made necessary to divide the emissions into two groups: (1) those mainly influenced by combustion technology and process conditions to compare between poor and high standard furnace design and (2) those mainly influenced by fuel properties.

Some related experiences were developed in the Netherlands [43–45], where different wood-burning installations with a capacity of 30–320 kW were working with the same fuel (clean woodchips), so that the technologies could be compared. It has to be noted that these installations are relatively efficient and that their automatic operation and combustion process control drastically reduced the emission of CO and C_xH_y [35]. In this respect, it is known that there are some design and operation parameters that help to reduce the gaseous emissions. Moreover, the correct dimensioning of boilers as well as the user behaviour may have an influence on the emissions.

References

[1] Larson, T.V. & Koenig, J.Q., Wood smoke: emissions and noncancer respiratory effects. *Annual Review of Public Health*, **15**, pp. 133–156, 1994.

[2] Hemminki, K. & Pershagen, G., Cancer risk of air pollution: epidemiological evidence. *Environmental Health Perspectives,* **102**(Suppl), pp. 187–192, 1994.

[3] Pope III, C.A., Lung cancer, cardiopulmonary mortality, and long-term exposure to fine particulate air pollution. *Journal of American Medical Association*, **287**, pp. 1132–1141, 2002.

[4] Andreae, M.O. & Merlet, P., Emission of trace gases and aerosols from biomass burning. *Global Biogeochemical Cycles*, **15**, pp. 955–966, 2001.

[5] Winiwarter, W., Kuhlbusch, T.A.J., Viana, M. & Hitzenberger, R., Quality considerations of European PM emission inventories. *Atmospheric Environment*, **43**, 3819–3828, 2009.

[6] ABB, *Emissions and the Global Environment – Statistics and Scenarios*, 1994.

[7] Karlsvik, E., Hustad, J.E., Skreiberg, Ø. & Sønju, O.K., Greenhouse gas and NO_x emissions from wood-stoves. In *Energy, Combustion and Environment*, Gordon & Breach, pp. 539–550.

[8] Magelli, F., Boucher, K., Bi, H.T., Melin, S. & Bonoli, A., An environmental impact assessment of exported wood pellets from Canada to Europe. *Biomass and Bioenergy*, **33**, pp. 434–441, 2009.

[9] Flyver Christiansen, H., Danish Energy Agency FMW., Center for Biomass Technology at dk-TEKNIK Energy and Environment, "LCA of Procurement and Conversion of Biomass and Fossil Fuels – Used for Energy Production in Denmark 1997," *Poster presented at the First World Conference and Exhibition on Biomass for En 2000.*

[10] Sikkema, R., Junginger, M., Pichler, W., Hayes, S. & Faaij, A.P.C., The international logistics of wood pellets for heating and power production in Europe: costs, energy-input and greenhouse gas balances of pellet consumption in Italy, Sweden and the Netherlands. *Biofuels, Bioproducts and Biorefining*, **4**, pp. 132–153, 2010.

[11] Thiers, S., Aoun, B. & Peuportier, B., Experimental characterization, modeling and simulation of a wood pellet micro-combined heat and power unit used as a heat source for a residential building. *Energy and Buildings*, **42**, pp. 896–903, 2010.

[12] Pelletsatlas, *English Handbook for Wood Pellet Combustion*, European Biomass Industry Association, 2009.

[13] Obernberger, I. & Brunner T., Aerosols in *Biomass Combustion – Formation, Characterisation, Behaviour, Analysis, Emissions, Health Effects*, vol. 6 of Thermal Biomass Utilization Series. BIOS Bioenergiesysteme: Graz, Austria, 2005.

[14] Gardiner, W.C., *Gas-Phase Combustion Chemistry*, 2nd edn., Springer-Verlag, 2000.

[15] Miller, J.A. & Bowman, C.T., Mechanism and modelling of nitrogen chemistry in combustion. *Progress in Energy and Combustion Science*, **15**, pp. 287–338, 1989.

[16] Bowman, C.T., Control of combustion-generated nitrogen oxide emissions: technology driven by regulation. *Proceedings of the 24th International Symposium on Combustion*, The Combustion Institute, Pittsburgh, PA, pp. 859–878, 1992.

[17] Zeldovich, Y.B., Barenblatt, G.I., Librovich, V.B., Makhviladze, G.M., *The Mathematical Theory of Combustion and Explosions*. Consultants Bureau: New York, pp. 30–36, 1985,.

[18] Zeldovich, Y.B., Sadovnikov, P.Y. & Frank-Kamenetskii, D.A., *Oxidation of Nitrogen in Combustion*. Academy of Sciences of the USSR: Moscow, 1947.

[19] Fenimore, C.P., Formation of nitric oxide in premixed hydrocarbon flames. *Proceedings of the Thirteenth Symposium (International) on Combustion*, The Combustion Institute, Pittsburgh, pp. 373–380, 1971.

[20] Jensen, L.S., Jannerup, H.E., Glarborg, P. & Jensen, A.D.-J.. Experimental investigation of NO from pulverised char combustion. *Proceedings of the Combustion Institute*, **28**, 2000.

[21] Johnsson, J.E., Formation and reduction of nitrogen oxides in fluidised bed combustion. *Fuel*, **73**, 1994.

[22] Skreiberg, Ø., Hustad, J.E. & Karlsvik, E., Empirical NO_x modelling and experimental results from wood-stove combustion. In *Developments in Thermochemical Biomass Conversion*, Blackie Academic & Professional, pp. 1462–1476, 1997,.

[23] Nussbaumer, T., *Furnace Design and Combustion Control to Reduce Emissions and Avoid Ash Slagging*. Report No. IEA Bioenergy: T13: Combustion, no. 03 1998, International Energy Agency, Biomass Combustion Activity, Final Report of the Triennium 1995–1997.

[24] Roy, M.M., Dutta, A. & Corscadden, K., An experimental study of combustion and emissions of biomass pellets in a prototype pellet furnace. *Applied Energy*, **108**, 298–307, 2013.

[25] Technology CFB., *Straw for Energy Production, Technology – Environment – Economy*, 2nd edn., www.videncenter.dk/, 1998.

[26] Houmøller, S. & Evald, A., Sulphur balances for biofuel combustion systems. *Presented at the 4th Biomass Conference of the Americas*, 1999.

[27] Vassilev, S.V., Baxter, D., Andersen, L.K. &, Vassileva, C.G., An overview of the chemical composition of biomass. *Fuel*, **89**, pp. 913–933, 2010.

[28] Said, N., Bishara, T., García-Maraver, A. & Zamorano, M., Effect of water washing on the thermal behavior of rice straw. *Waste Management*, **33**, 2250–2256, 2013.

[29] Win, K.M., Persson, T. & Bales, C., Particles and gaseous emissions from realistic operation of residential wood pellet heating systems. *Atmospheric Environment*, **59**, 320–327, 2012.

[30] Nussbaumer, T., Emissions from biomass combustion, IEA Biomass Agreement, Task X – Biomass Utilization, Activity 1: Combustion, Final Report of the Triennium 1992–1994, 1994.

[31] European Union, Directive 2001/80/EC, PB L 309, Guidelines for Large Combustion Plants (LCP), n.d.

[32] European Union, 2000/76/EG, PB L 332, Guidelines for Waste Incineration Directive (WID), n.d.

[33] EN303-5, Heating boilers for solid fuels, manually and automatically stoked, nominal heat output of up to 500 kW. Terminology, requirements, testing and marking, 2012.

[34] Verma, V.K., Bram, S., Vandendael, I., Laha, P., Hubin, A. & De Ruyck, J., Residential pellet boilers in Belgium: standard laboratory and real life performance with respect to European standard and quality labels. *Applied Energy*, **88**, pp. 2628–2634, 2011.

[35] Van Loo, S. & Koppejan, J., *Biomass Combustion and Co-firing*, Earthscan: London and Sterling (VA), 2004.

[36] Skreiberg, Ø., Theoretical and experimental studies on emissions from wood combustion. *PhD Thesis*, Norwegian University of Science and Technology, 1997.

[37] Skreiberg, Ø., Karlsvik, E., Hustad, J.E. & Sønju, O.K., Round robin test of a wood-stove: the influence of standards, test procedures and calculation procedures on the emission level. *Biomass and Bioenergy*, **12**, pp. 439–452, 1997.

[38] Skreiberg, Ø. & Saanum, Ø., Comparison of emission levels of different air pollution components from various biomass combustion installations in the IEA Countries. Supplemental Report to IEA Report Emissions from Biomass Combustion, IEA Biomass Agreement, Task X – Biomass Utilizatio, 1994.

[39] Karlsvik, E., Hustad, J.E. & Sønju, O.K., Emissions from wood-stoves and fireplaces. In *Advances in Thermochemical Biomass Conversion*, Blackie Academic & Professional, pp. 690–707, 1993.

[40] Strehler, A., *Emissionsverhalten von Feurungsanlagen fur feste Brennstoffe*, Technische Universitat Munchen-Weihenstephan, Bayerische Landesanstalt fur Landtechnik, Freising, 1994.

[41] Nussbaumer, T. & Hustad, J.E., Overview of biomass combustion. In *Developments in Thermochemical Biomass Conversion*. Blackie Academic & Professional, pp. 1229–1243, 1997.
[42] Obernberger, I., Stand und Entwicklung der Verbrennungstechnik. VDI Berichte 1319, *Thermische Biomassenutzung, Technik und Realisierung*, April, VDI Verlag GmbH: Düsseldorf, Germany, 1997.
[43] Landentechnik, P.B.F., Wieselberg/Erlauf Prot. Nr. 033/94, 32/94,024/95, 025/94, 035/95, 023/93, n.d.
[44] Wildeburger, J.M.F., Staatlich befugter und beeideter Zivilingenieure für Technische Chemie, Polten (Aus), Prüfberichte PZ 1304/ZJW 920, 1996.
[45] Sulilatu, W.F., *et al.*, Kleinschalige verbranding van schoon resthout in Nederland, NOVEM rapport, November, 1992.

CHAPTER 7

Particulate Emissions from the Combustion of Biomass Pellets

M. Rabaçal & M. Costa
Mechanical Engineering Department, Instituto Superior Técnico, University of Lisbon, Lisbon, Portugal.

Abstract

The combustion of biomass pellets in small-scale appliances for domestic heating is a common source of particulate matter (PM) emissions, especially in Europe. The diversification of biomass raw material associated with the growing pellet market has the potential to increase PM emissions, particularly fine inorganic particles. Global concerns on public health and environmental impacts have imposed limitations to the emission of particles creating a need for a better understanding of the formation of particles to develop appropriate mitigation strategies. This review addresses the quantification, formation and mitigation of PM emissions. The measurements of PM emissions can be considerably affected by sampling and measurement methods or instruments used; thus, there is a need for uniform standards and protocols. PM emissions from small-scale pellet boilers have been related to both fuel quality and operating conditions. Biomass characteristics, specifically the ash content and the ash composition, have a significant impact on PM emissions as the ashes of the fuel contain the precursors for the formation of inorganic particulate matter. Operating conditions have also an important impact as soot and organic particles are formed in poorly oxygenated environments. Mass and number concentrations, mass and number size distributions, and chemical compositions of PM emissions are reported from the most relevant related studies. On-going research on mitigation strategies focuses on the inhibition of PM formation by combustion optimization and fuel pre-treatment; or on the separation of particles in the flue gas.

Keywords: Biomass pellets, small-scale boilers, particulate matter, sampling, measurements, formation, emissions.

1 Introduction

Although small-scale pellet appliances have reached a level of maturity that competes successfully with traditional oil and gas heating devices, they still pose issues related to the emission of particulate matter (PM). PM emissions from modern pellet appliances under typical operations are far lower than manual stoves; however, variations in the operation due to variable loads may lead to poor combustion conditions and lead to higher PM emissions [1]. Furthermore, the pellets market is growing and the production of second grade quality pellets with higher ash content is expected to increase, contributing to a potential increase of inorganic PM emissions [2]. PM emissions may also influence the performance of the appliances as a fraction of these deposits on the combustion chamber and heat exchanger surfaces causing operational problems [3]. Finally, there is a worldwide growing concern on the exposure of the population to fine particles, which can lead to health issues [4] and negative impacts on the environment [5].

Particles formed during combustion are generally classified into two categories: primary and secondary particles. Primary particles are formed at high temperatures in the combustion zone, whereas secondary particles are formed in the exhaust plume and the atmosphere. In small-scale pellet burner appliances, there are three main types of primary particles – soot particles, inorganic particles and organic particles – and the formation of those varies with combustion conditions, fuel properties and type of appliances [6–9]. Soot and organic particles originate from combustible material, whereas inorganic particles are originated from the ash content of the fuel. A considerable amount of research on this subject has been done in the past [10–28]. Studies have covered both the fundamental issues, at a laboratory scale using reactors with very well-defined and controlled conditions [22–28], and applied issues using commercial appliances either under laboratory controlled operation of real-life operation [10–21].

An attempt is made here to present a summary of the most relevant studies on particulate matter emissions from the combustion of biomass pellets in small-scale boilers covering a broader range of related aspects, complementing previous reviews [3,29]. This chapter is presented as follows. First, the legislative framework is presented, where the most relevant standards for PM emission limits in the European Union (EU) are presented. Then, sampling and measuring methods are listed, including the most relevant standards and a description of the state-of-the-art instruments available. The formation of particles follows, where a description of the mechanisms behind the formation of particles is described and particle dynamics are reviewed. Finally, a review on PM emissions from pellets appliances is presented, including an overview of relevant studies, data report concerning how operating conditions and fuel composition affect PM emissions and current state-of-the-art mitigation strategies.

2 Legislative Framework

Combustion is the most mature technology for biomass utilization, but emissions from biomass combustion can contribute significantly to concentrations of particulate matter in ambient air [30]. Wood burning is one of the major contributors of primary particle to the atmosphere during winter times over large parts of Europe [31–34], and small-scale combustion systems in particular play an important role [31,34].

Residential combustion of biofuel is carried out in simple wood stoves as well as in well-controlled devices, such as pellet burners and modern wood boilers, and the emissions of pollutants vary over a wide range depending on the technology of the combustion furnace and on the quality of the combustion and biofuels used [10,35,36]. Nevertheless, it is well known that in general residential heating is an important local source of ambient air particles [31–34]. This results in negative health effects especially in densely populated areas, where these systems can cause high concentration of particulate matter in the ambient air [11,37]. In contrast to the large amount of information relating urban PM to human health impacts, there are only a limited number of studies directly evaluating the community health impacts of air pollution resulting from the burning of biomass [4]. The dust emissions can consist of carbonaceous particles and vaporized inorganic matter mainly alkaline metals, sulphur and chlorine [36], and its inhalation may result in respiratory distress, asthma, cardiovascular illness and even mortality [4].

These factors make evident the necessity to develop governmental regulations to ensure a safe, environmental-friendly and healthy operation of boilers. In this regard, the European Standard EN 303-5 [38] governs throughout the EU. The EU standard identifies three quality classes of boilers, based on efficiency and emission requirements (Fig. 1).

Depending on its class and also on its feeding system, the European Standard establishes different thresholds for the PM emissions from biomass boilers (Table 1). It is noted, however, that countries such as Germany, Austria and Switzerland have national standards that are stricter in terms of dust emissions (Table 1), efficiency and safety requirements than the EU regulations [39].

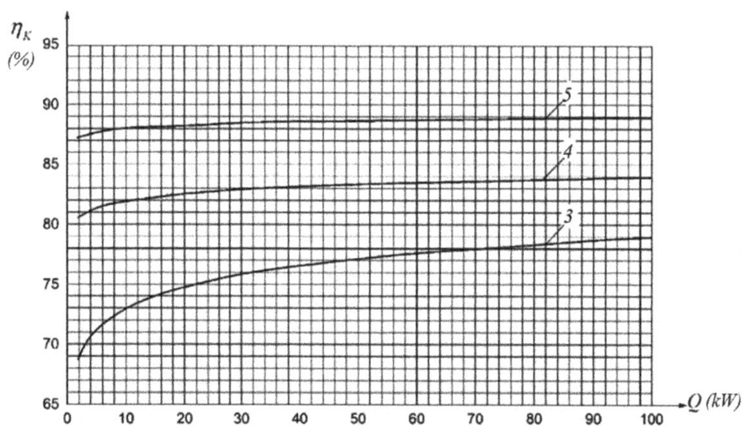

Figure 1: Classification of boilers according to their nominal power (Q) and efficiency (η_K) [38].

Table 1: PM emissions thresholds for biomass boilers <50 kW [38,40].

Type of boiler	PM emissions thresholds (mg/m^3 10% O_2*)					
	EN 303-5			German Ministry of Economic Affairs and Technology	Swedish Boverket	Austrian BimSch
	Class 3†	Class 4	Class 5			
Manually fed		75	60			
Automatically fed	150	60	40	50	100	150

*Dry flue gas, 0°C, 1013 mbar.
†Non-woody biomass fuels do not have to fulfil these limits, but their PM emissions have to be indicated and do not exceed 200 mg/m^3 10% O_2.

Table 2: PM emissions thresholds for biomass boilers <50 kW considering the European ecolabels [39].

European Ecolabels	PM emissions thresholds (mg/m^3 10% O_2*)
Swan	70
Flamme Verte	165
Blue Angel	30
UZ-37	15
SEI	150

*Dry flue gas, 0°C, 1013 mbar.

In Germany, these standards are given by 'German Ministry of Economic Affairs and Technology,' being the most restrictive regulation in Europe in terms of permissible PM emission values. In Sweden, the use of pellet boilers and pellets fuel for household heating purposes is regulated by Swedish regulations given by 'The National Board of Housing, Building and Planning' (Boverket) and 'The Swedish Standards Institute' (SIS). Boverket recommends limit values for small scale biomass combustions only, whereas SIS requires their compliance with the values given by European Standard EN-303-5, which has since 1999 been applied as Swedish Standard [39]. In Austria, BimSch (Federal Low for Emission Protection-Austria) regulates the emissions from pellet burning appliances in addition to Ö-Norms-EN-303-5 as an equivalent EU regulation [39].

In addition to these governmental regulations, some European countries developed stricter regulations in the form of ecolabels. This is the case of 'Swan' in the Scandinavian countries, 'Flamme Verte' in France, 'Blue Angel' in Germany, 'UZ-37' in Austria or SEI in Ireland (Table 2).

3 Sampling and Measuring Techniques

The sampling and measurement of particles emitted from pellet boilers and can be a very challenging task. Particles can be sampled from the combustion chamber of the boiler or in the chimney/stack, depending on the aim of the measurement – boiler testing, health and environment impact studies or particle formation studies. Particle concentration measurements may be affected by temperature, pressure, dilution and even by losses during the sampling procedure through inertial deposition or collision with the walls [41]. Although there is a variety of measurement procedures, a common feature of the vast majority is that the measurement is done from filter-based time-integrated samples [42]. A number of standards have been developed to establish quality references for PM measurements, to list a few: SS028426 (1991-12-04) [43], SFS 3866 (1990-12-11) [44], NS 3058-2 [45], British Standard 6434:1969 [46], VDI 2066 Blatt 2 [47], VDI 2066 Blatt 5 [48], EN 13284-1 [49], MEL 02 [50], ISO 9096 [51], ISO 10155 [52] and ASTME 2515 [53]. These standards consider different sampling locations, different conditions and capture techniques. Some of the aforementioned standards are summarized in Table 3. For the sake of clarity, we define *sampling* as the act of collecting particles from a target location and *capture* as the act of capturing the sampled particles in a plate/filter. The samples are measured during the collection itself or post-processed. The suction probe is the most established sampling technique; whereas sampling location and capture techniques may vary between standards, see Table 3.

When the aim of a measuring campaign is to assess the level of emissions of a boiler, the sampling can be done either in the chimney, in a dilution tunnel or in a dilution hood. Since within the chimney the conditions are above the dew point of the flue gas, the dilution tunnel and the dilution hood methods were established to give a good estimation of the contribution to ambient level of particles [41] (see Figs 2 and 3). In the dilution tunnel, the fast cooling of the flue gas, due to the entrainment of ambient air in the dilution tunnel, will lead to the condensation of particle precursors into the solid phase increasing the concentration of particles measured. Measurements at the chimney are more affordable as they require less equipment [41]. This effect can be reproduced in chimney sampling as well when a dilution section is added to the sampling train, typically in the sampling probe itself.

For measurements inside the combustion chamber, there are no standards established. These types of measurements are designed to study the formation of the particles and the evolution of the particle properties in the chamber [22]. In principle, any type of capture technique may be used, provided that the sampling probe is capable of withstanding typical combustion temperatures and slagging. However, the cascade impactor or coupling cyclone separators upstream are preferred in detriment of using simpler filters. The cascade impactor method allows separating the sample in different size classes, which can be further chemically analyzed giving a broad view of the particle physical and chemical state at the measurement location. An installation for the measurement of PM from inside the

Table 3: Summary of standards for the measurement of PM emissions [45,48–50].

Standard	EN 13284-1	VDI 2066 Blatt 5	NS 3058	MEL 02
Sampling location	In-stack or out-stack	In-stack	Dilution tunnel	In-stack or out-stack
Sampling velocity	Iso-kinetic	Iso-kinetic	Measured with pitot	
Sampling technique	Suction probe	Suction probe	Suction probe	Suction probe
Sampling pressure (mBar)	–	600	–	
Capture technique	Plate filter (paper, gas fibre or PTFE)	Cascade impactor (metal plate)	Plane filter (1 μm porosity)	Plane filter (quartz)
Flow rate (m^3/h)	1–3 in-stack 1–20 out-stack	4 or 12	–	–
Concentration range (mg/Nm^3)	<20	1–1000	~1600 at 13% O_2	50
Particle size (μm)	Total	0.5–15	Total	Total
Backup filter	No	Yes	No	No

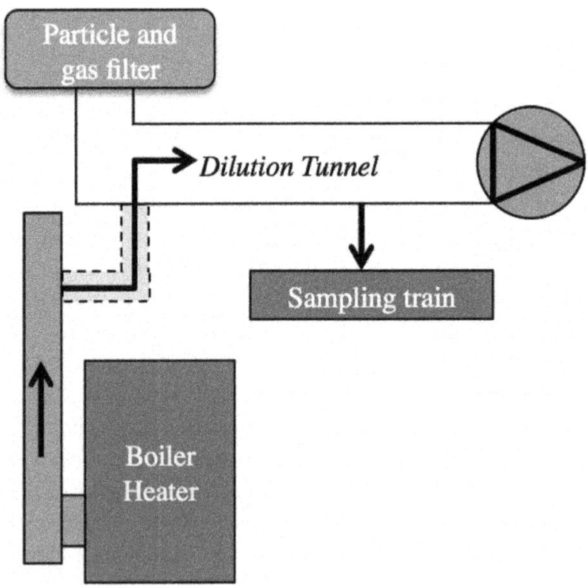

Figure 2: Diagram of an installation for PM measurements using a dilution tunnel.

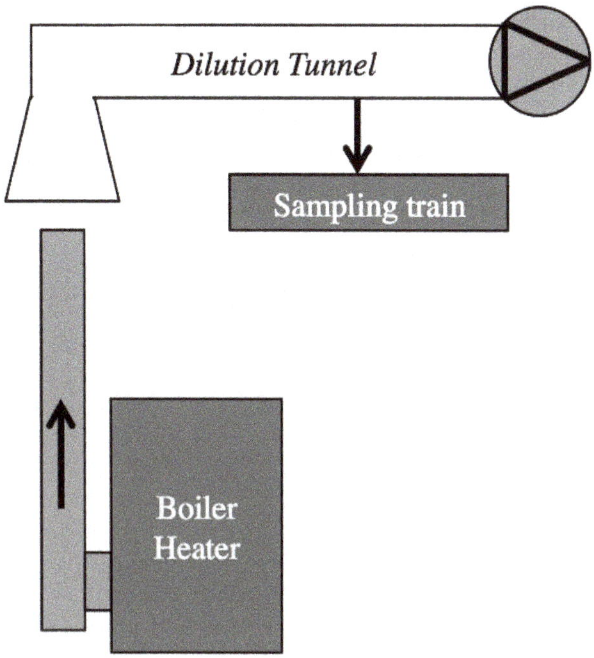

Figure 3: Diagram of an installation for PM measurements using a dilution hood.

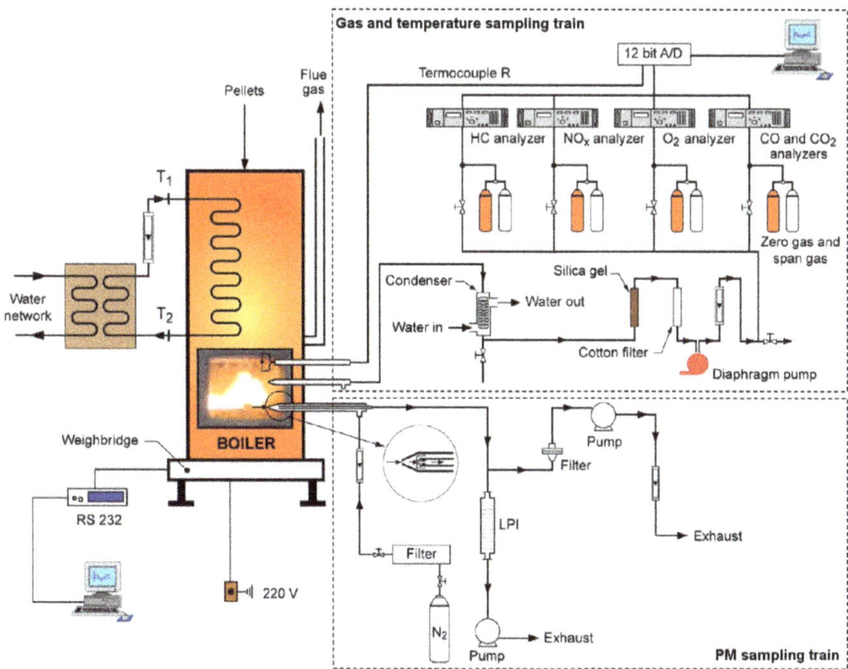

Figure 4: Schematic of an experimental setup for the study of particle formation inside the combustion chamber of a pellets boiler [22].

combustion chamber is exemplified in Fig. 4. Details of the instruments and procedures can be found in the work by Fernandes and Costa [22]. The type of probe and its operation may introduce different types of artefacts to the measurements. Sampling may be done by aerodynamic-quenching particle sampling (AQPS) probes, dilutions probes and/or thermophoretic sampling probes [54]. In the first two types of probes, gas with entrained particles is drawn into the sampling train by suction. In the latter type, a plate holding a filter is inserted in the hot flow for a short period of time and the particles subjected to the temperature gradient move from the high temperature region to the low temperature region and, finally, depositing in the filter (usually a TEM grid). Artefacts arise mainly due to the condensation of vapor precursors into the solid phase, although in AQPS probes these are less significant [54].

Generally, PM samples may be characterized in terms of mass concentration, mass size distribution, number concentration, number size distribution and chemical composition, each defined as follows:

- *Mass concentration:* Mass of particles per unit of volume of flue gas (mg/Nm^3).
- *Mass size distribution:* Distribution of relative amounts of particles by mass in size intervals over a size range (mg/Nm^3).
- *Number concentration:* Number of particles per unit of volume of flue gas (1/Nm^3).
- *Number size distribution:* Distribution of the relative number of particles in size intervals over a size range (1/Nm^3).

The selection of a capture technique will define the type of measurements that can be done, and the advantages and disadvantages of the different techniques have to be weighted. Table 4 shows a summary of the commonly used state-of-the-art PM collecting instruments. When designing the sample train, one has to take in consideration the different operating conditions of each instruments (flow rate and pressure). In some cases, a cyclone has to be placed upstream of the instrument when a size cutting is intended. Cascade impactor instruments provide size-distributed samples.

Figure 5 shows a comparison of number size distributions obtained with different measurement instruments at simultaneous measurement in flue gas from a residential pellet burner by Pagels *et al.* [56]. In the submicron range (<0.5 μm), ELPI, DLPI and SMPS measurements show a reasonable agreement. Above a particle size of 0.5 μm, DLPI and APS measurements show a similar trend of a steep decrease of concentration of larger particles, in contrast to ELPI. Concentrations differ in several orders of magnitude – the highest concentrations are measured by ELPI, followed by DLPI and APS. During successive separation in the ELPI, a minor carryover of small particles are transported from smaller size sections to larger size sections, which can lead to an overestimation of the number of larger particles [1].

Finally, the samples collected may be chemically analyzed providing information on combustion conditions and which ash elements populate the PM emissions. For this type of analysis, it is interesting to use size distributed samples as

Table 4: Summary of PM sample collecting techniques [3,13,30,55].

	Filter sample	DLPI	ELPI/ ELPI+	APSS	SMPS	FMPS	TEOM	SEM/TEM	DENUDER
Size range	Depends on pre-utting	30 nm–10 µm	7 nm–10 µm/ 6 nm–10 µm	50 nm–20 µm	2.5 nm–1 µm	5.6 nm–560 µm	Depends on pre-cutting	–	Depends on pre-cutting
No. of size classes	–	12	12/14	–	–	16	–	–	–
Sample flow rate (lpm)	–	10 or 30	10 or 30/10	1	0.2–2	10	Main: 3 Bypass: 14	–	–
Capture plate diameter (mm)	–	25	25	–	–	–	–	–	–
Pressure (mbar)	–	100	100/40	400–1030	750–1050	700–1034	–	–	–
Weight (kg)	–	5	35/15	10	23.2	32	18	–	–
Measures	–	Aerodynamic diameter	Aerodynamic diameter	Aerodynamic diameter	Mobility diameter	Mobility diameter	–	–	–
Measurement method	Continuous	Continuous	Continuous	Continuous	Continuous	Continuous	Discontinuous	Discontinuous	Discontinuous
Parameters		Mass size distribution	Number concentration and mass distribution	Number concentration and mass distribution	Number concentration and mass distribution	Number concentration and mass distribution	Mass size distribution	Morphology, particle size, elements can be analyzed	Gas-particle distribution of organic compounds
Advantages	Simple and reliable	Large size range	Robust; real-time; large size range	Larger particles	Very smaller particles	Fast; indicate changes in process well	Agree well with filter samples	–	No secondary reactions
Disadvantages	No size range	Time-integrated	Results may be affected due to wide chan-nels plate	Not suitable for smaller particles	Scanning time	More inaccurate than SMPS	Filters need to be replaced if concentration is high	Sensitive to amount of sample	Effort; particle losses

DLPI, Dekati low pressure impactor; ELPI, electrostatic low pressure impactor; APSS, aerodynamic particle size spectrometer; SMPS, scanning mobility particle sizer; FMPS, first mobility particle sizer; TEOM, tapered element oscillating micro-balance; SEM/TEM, scanning electron microscope/transmission electron microscopy.

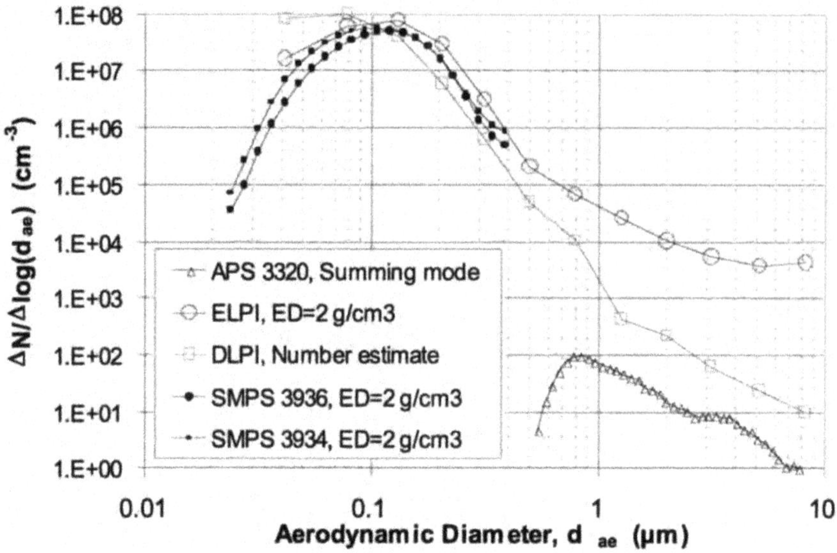

Figure 5: Comparison of number size distributions with different measurement instruments at simultaneous measurement in flue gas from a residential pellet burner [56].

particle composition and particle size are correlated. Chemical analyses may be conducted using the following methods [3]:

- Energy dispersive spectroscopy (EDS).
- Energy dispersive X-ray spectroscopy (EDX).
- Environmental scanning electron microscopy (ESEM).
- Inductively coupled plasma-mass spectrometry.
- Inductively coupled plasma-optical emission spectrometry (ICP-OES).
- Particle induced X-ray emissions (PIXE).
- Scanning electron microscopy (SEM).
- Transmission electron microscopy (TEM).
- Time of flight-secondary ion mass spectrometry.

4 Particle Matter Formation in Combustion Systems

4.1 Inorganic particle formation

Biomass has a considerable amount of inorganic material that is the source of the inorganic particles formed during combustion. The inherent inorganic material is mostly composed by simple inorganic salts (oxides and hydroxides of silicon, nitrates, sulphates, chlorides, phosphates and oxalates of the alkali and alkaline earth metals Na, K, Mg and Ca) [57]. Extraneous inorganic material is added to the biomass through geological processes, during harvesting, handling and processing of the fuel. Generally, extraneous material includes sand, soil or other mineral materials. In laboratory characterization, the most common elements are standardly expressed as oxides, in their highest oxidation state, i.e. SiO_2, Al_2O_3, Fe_2O_3, CaO, MgO, TiO_2, Na_2O, K_2O, P_2O_5, SO_3 and ZnO [58]. The inorganic particle forming elements can furthermore be divided into three categories: (1) non-volatile elements, Si, Al, Fe, Ca, etc., (2) highly volatile elements, Na, K, Cl, S, etc. and (3) highly volatile heavy metals, Zn, Cd, Pb, etc.

The process starts in the biomass particle where the inorganic salts are decomposed. Due to their non-volatile nature, elements such as Ca, Mg, Si, Al and Fe are typically found in coarse fly ash particles [14,23,25,26,59,60]. The feeding system (top, horizontal or bottom) has an important influence on coarse fly ashes emissions, as disturbances in the fuel bed cause fragmentation of char pellets and the fragments particles are entrained in the fluid flow as mentioned above [10,15,16]. Highly volatile elements, in particular Na and K, are dominant species in the fine particle fraction [15,17,18,22,35,60–62]; hence, a major interest in the study of fine inorganic particle formation has been the release of alkali species.

There are few studies on alkali release in commercial domestic pellet burners operating under realistic conditions, but a number of studies performed on laboratory-controlled conditions have shown that a part of the organically bond alkalis are released during the lower temperature stage of devolatilization of biomass and the major fraction of alkali release occurs during the higher temperature stage of char combustion [63–65]. In fact, temperature has significant effect on alkali release and on the emission of fine particles [17,19,28,60,66]. Given that the presence of moisture in the fuel lowers the temperature in the fuel bed, due to evaporation, this parameter may have an impact on alkali release [67]. Finally, the chemical composition of the fuel ash has a strong effect on the release of alkali metals [25,28,68]. Since most modern domestic pellet burners operate at similar conditions, the fuel ash composition has the most pronounced impact on fine PM emissions [16,18,20,21,62].

The presence of Si can reduce the release of alkali metals to the gaseous phase when these react to form non-volatile alkali silicates and ultimately reduce fine particle emissions [19,20,28,69]. The addition of Kaolin ($Al_2Si_2O_5(OH)_4$) was shown to have a significant role in reducing fine particle emissions through the global reaction rates shown in eqns (1) and (2) [15]. It was observed an increase in HCl concentration in the gaseous phase together with an increased fraction of K and P in the coarse fly ash and in the bottom ash, as a consequence. The presence of sulphur may also contribute to the retention of alkaline metals, as it will be discussed later.

$$Al_2Si_2O_5(OH)_5 + 2KCl \leftrightarrow 2KAlSiO_4 + H_2O + 2HCl \quad (1)$$

$$Al_2Si_2O_5(OH)_5 + 2KCl + 2SiO_2 \leftrightarrow 2KAlSi_2O_6 + H_2O + 2HCl \quad (2)$$

Nucleation of Zn has been suggested to form the first fine fly ash particles in the flue gas when ZnO is present in the fuel ash [25,28,70,71]. However, in the presence of more complex solid zinc than pure zinc oxide or some simple zinc salt, the formation mechanism and the exact phase composition of zinc containing particles still remains to be elucidated [28].

It has been postulated that K and Na are released from the biomass particle mainly as highly volatile chloride vapors (KCl and NaCl) and hydroxide vapors (KOH and NaOH) [6–9]. Alkali release in the form of chloride is associated with the degree of release of the highly reactive HCl [19,25,28]. The release of HCl may act as a limiting or a promoting factor. It has been shown that the addition of HCl to biomass combustion increases the release of alkali metals [19]. On the other hand, in low-HCl conditions alkali species tend to react with silicates [72].

Sulphation of alkali metals can be described by the global reactions in eqns (3) and (4), where A can be K or Na. During the cooling phase, alkali sulphates are estimated to nucleate or condense on existing seed particles soon after their formation (~950°C), much before alkali chlorides nucleate or condense (~600°C) [9,28,54,73].

$$2AOH + SO_2 + \frac{1}{2}O_2 \leftrightarrow A_2SO_4 + H_2O \quad (3)$$

$$2ACl + H_2O + SO_2 + \frac{1}{2}O_2 \leftrightarrow A_2SO_4 + 2HCl \quad (4)$$

In some conditions, it has been shown that the addition of SO_2 in HCl-rich combustion leads to sulphation and condensation in the fuel particle site, decreasing the release of alkali species to the gaseous phase [17,19,69]. Contrary to these findings, it was also observed that in HCl-rich conditions the addition of SO_2 promoted the conversion of gaseous chlorines to sulphates via reaction in eqn (4) leading to an increase of sulphates in the fine particle fraction. It is important to point out that a kinetic limitation to the reactions in eqns (3) and (4) was found for reactions of K [54] for temperatures below 820°C, which may have had an influence in the findings above. SO_2 is a precursor of K_2SO_4; therefore, at temperatures lower than 820°C the rate of production of K_2SO_4 is limited by the oxidation of SO_2 to SO_3, as described by the reactions in eqns (5) and (6) [54]

$$SO_2 + O + M \leftrightarrow SO_3 + M \quad (5)$$

$$SO_3 + O \leftrightarrow SO_2 + O_2 \quad (6)$$

If the flue gas contains more alkaline metals than what is available for bounding with S and Cl, then the formation of alkali carbonates will occur [19]. A limitation in the carbonate formation may lead to the formation of nitrates [13]. These bounds can be better understood resorting to the molar ratios (Na+K)/Cl and (Na+K)(Cl+2S) of the emitted particles [9,60,72]. The (Na + K)/Cl ratio is unity if all the alkali species are bound as chlorides, and the (Na + K)/(Cl + 2S) ratio is unity if all alkali metals are bound as sulphates and chlorides. In turn, if both ratios are equal to unity, all the Cl and S are bound with alkali metals. If the second ratio is still higher than unity, alkali species were bound as other compounds, such as carbonates or nitrates, in addition to chlorides and sulphates.

The emitted fine ash fraction is typically formed of alkali sulphates, chlorides and carbonates, as shown above, but in phosphorus-rich fuels alkali phosphates have been found to be one of the major compounds in the fine particle emissions [19,28]; however, the role of phosphorus is still poorly understood [9].

4.2 Soot and organic particles formation

Soot particles are a complex mixture of amorphous elemental carbon and organic material that are formed during devolatilization and volatile combustion. Although soot has been extensively studied in coal combustion, there are only a few studies on biomass combustion and the complex formation mechanism is currently not fully understood [3,6,8,9,26]. Precursor molecules are formed in the fuel-rich area when hydrocarbon fragments collide with each other and react with surrounding gases to form polycyclic aromatic rings (PAH) rather than being oxidized. PAH then grow by polymerization up to a molecular mass between 500 and 2000 a.m.u and nucleation occurs forming soot nuclei with a size of 1–2 nm [6,8]. Thereafter, the soot nuclei grow by surface reactions, agglomeration and condensation. In non-premixed flames, where even in excess air conditions fuel-rich eddies are formed, soot if formed within the flame and is oxidized as it is convected to the boundary of the flame [6,9]. Fernandes and Costa [22] observed that, in the combustion chamber of a pellets boiler, soot particles are mainly formed early in the combustion process and tend to oxidize rapidly in regions of high temperature and high oxygen availability.

During biomass devolatization, a large variety of organic compounds with varying molecular weights and vapor pressures is released into the gas phase forming the volatile gas [3,9]. These include mainly levoglucosan, aliphatic and oxygenated compounds, terpenoids, resins, gum and other biopolymers [9]. If combustion conditions are poor, these elements will be released into the fuel gas. They are typically divided according to their boiling points into very volatile organic compound (VVOC), volatile (VOC) and semivolatile organic carbon (SVOC) compounds [3,8], although VOC is a term generally used that encompasses the three types. Following the cooling of the flue gas, typically at temperatures lower than 500°C, the VOC will condense [3]. VOCs condense mainly on exiting particles in the flue gas; however, with dilution and rapid cooling, VOCs may nucleate homogeneously [9].

4.3 Particle morphology

Particle morphology and chemical analysis allow characterizing PM emissions and it is also a quite useful tool to understand particle formation and dynamics [14,17–19,22,26,61,72]. Figure 6 shows SEM images of PM samples collected on quartz microfibre filters, showing different types of PM structures emitted during the combustion of pine pellets in a domestic pellets boiler [14]. It is possible to identify three types of PM structures, namely: (1) ultrafine (<100 nm) particles (Fig. 6a), (2) sub-micrometre (Fig. 6a) and micrometre (Fig. 6b) sized particles/agglomerates and (3) isolated spherical and irregularly shaped large particles (Fig. 6c and d). Figure 7 shows the SEM images of PM samples collected on different substrates of a low-pressure impactor and the corresponding chemical composition, emitted during the combustion of olive pruning in a domestic pellets boiler [18]. Ultrafine particles are dominated by K and Na (Fig. 7a and b), whereas micrometre sized particles are dominated by Ca (Fig. 7c).

Micrometre sized particles are typically dominated by Si, Al, Ca and to a less extent Mg, Mn and K and present various morphologies, in particular isolated spherical particles, cubic or hexahedral particles with sharp edges (crystalline), and irregular shape (skeletal or agglomerates of very small particles) [14,17–19,22,26]. Literature indicates that micrometre sized particles are mainly originated from the entrainment of ash fragments due to char break up in the fuel bed. The presence of spherical particles (Fig. 6c) indicates that ash inclusions in the biomass char will melt and form droplets under the combustion [26,54]. Calcium-rich particles may

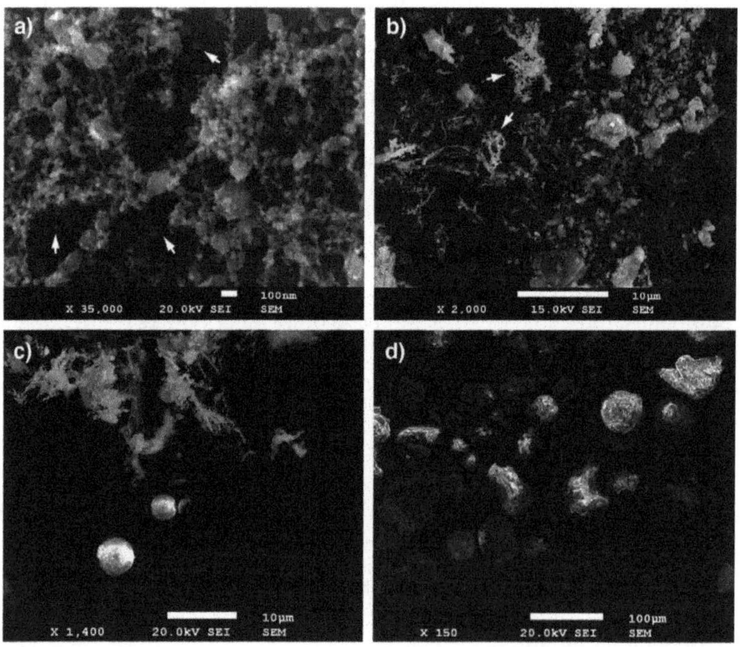

Figure 6: SEM images of PM samples collected on the quartz microfibre filters, showing different types of PM structures. (a) Sub-micrometre sized particles/agglomerates; (b) micrometre sized particles/agglomerates; (c) separate spherical micrometre sized particles; (d) separate irregularly shaped micrometre sized particles [14].

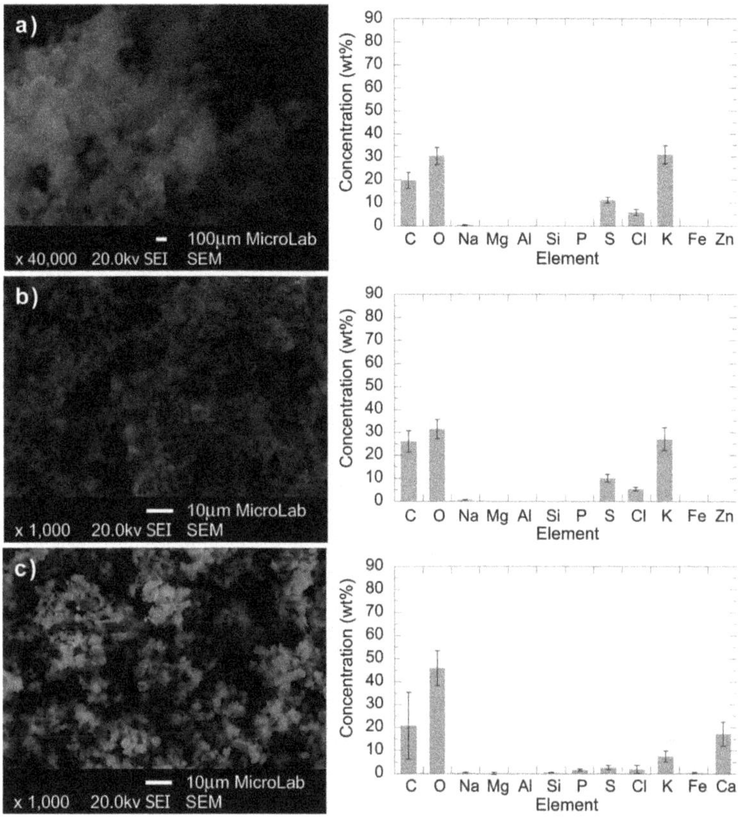

Figure 7: SEM images of PM samples collected on different substrates of a low-pressure impactor and the corresponding chemical composition, emitted during the combustion of olive pruning in a domestic pellets boiler. (a) Under 0.158 μm; (b) under 0.95 μm and (c) under 2.979 μm [18].

present distinctive morphologies [26,61]. Calcium-rich particles with a cubic crystalline and like a grain may consist of molten ash as well, as different temperature histories in the melting or solidification phases create different shapes [26,61]. Irregular agglomerates are typically composed of unburned carbon and non-volatile elements and to some extent volatile elements, indicating that ash inclusions within the biomass char agglomerate in the fuel bed during combustion followed by heterogeneous condensation alkali chlorides and sulphates on the surface [17,18,22,72].

Fine submicron particles are dominated by carbon, alkali species, chlorine and sulphur [14,17–19]. The origin of alkali species, chlorine and sulphur, is discussed in detail in the previous chapter. The presence of carbon indicates that soot was formed during combustion. Soot consists mainly of amorphous elemental carbon and organic material and soot particles are formed mainly in rich areas of the flame.

4.4 Particle dynamics

Particle formation and transport processes are described by the General Dynamic Equation (GDE), eqn (7) which governs the particle population balance in a control volume [74–77]. Here, ϕ represents the particle size distribution considered as a discrete size distribution with J modal points.

$$\frac{\partial \varphi}{\partial t} + \left(\nabla \times \varphi\left(v_{gas} - v_{p}\right)\right) = \nabla \times D_{p}\rho_{gas}\frac{\varphi}{\rho_{p}} + S_{nucl} + S_{cond} + S_{coag} + S_{\varphi} \quad (7)$$

The rate of change of ϕ over time in a control volume (first term LHS) is determined by the convective transport due to gas velocity, v_{gas} (ms^{-1}), and the particle velocity, v_p (ms^{-1}), (second term LHS), particle diffusion into and out of the control volume controlled by the diffusion rate, D_p (m^2s^{-1}), (first term RHS) and source terms due to nucleation, condensation and coagulation (second, third and fourth terms RHS, respectively). These source terms will be described in the subsections below. The last source term in the RHS accounts for fragmentation, inertial impaction and diffusion to surfaces. Although these phenomena have a relevant importance for the quantification of PM emissions, they are not directly relevant to the inorganic PM formation mechanisms. The reader is referred to [74–77] for more detailed information. Although the description of the governing equations of the gas and the particle velocities is out of the scope of this review, it is important to note that due to the extremely small dimensions of fine particles, themophoretic and turbophoretic forces may have a significant effect on the governing of the particle velocity [74–77].

The phenomena addressed in Section 1.2 are only fully described when considering the mass conservation balances of the involved species. The mass conservation equation of the mass fraction of species i present in the gaseous phase is described as follows:

$$\frac{\partial \rho Y_i}{\partial t} + \nabla \times Y_i v_{gas} = \nabla \times \left(D_i Y_i\right) + \omega_i + S_{phase} \quad (8)$$

The rate of change of the mass fraction Y_i (–) over time in a control volume (first term LHS) is determined by the convective transport due to gas velocity, v_{gas} (second term LHS), diffusion controlled by the diffusion rate, D_i (m^2s^{-1}), (first term RHS), formation or consumption controlled by the reaction rate, ω_i ($kg\ s^{-1}$) (second term RHS) and a source term, S_{phase}, due to phase change, i.e. nucleation and condensation (third term RHS). Again, the source terms describing nucleation and condensation will be addressed in the following subsections.

The reaction rates of the inorganic species involved are defined in reaction mechanisms, in which complexity goes far behind the scope of this review. FactSage© provides a series of reaction mechanisms and thermodynamic properties databases, allowing to perform thermodynamic equilibrium calculations under a prescribed temperature history, which have been used to better understand

PM formation [17,27,28,54,59,71]. Note that these calculations do not include kinetic limitations, such as the one described for sulphation of K described in Section 1.1.

4.4.1 Nucleation

Nucleation is a gas-to-particle process affecting the total mass of particles and total number of particles. The alkali vapor phase consists of monomeric molecules and clusters. Clusters may grow by molecular collision or shrink due to the weak surface forces, which are not capable of retaining the molecules [74–77]. When the partial pressure of a vapor equals its saturation vapor pressure, the cluster reaches mass equilibrium, as the deposition rate of molecules equals the release rate of molecules. This is known as the Kelvin effect and the modified saturation vapor pressure p_{sat} (Pa) is defined as follows [74–77]:

$$p_{sat} = p_s \exp\left(\frac{4\sigma V_i}{d_p k_B T}\right) \tag{9}$$

and the saturation ration S (–) is defined as:

$$S = \frac{p}{p_s} \tag{10}$$

where p_s (Pa) is above a flat surface, V_i (m^3) is the volume of species i, σ ($J\ m^{-2}$) is the surface tension, d_p (m) is the cluster diameter, k_B (JK^{-1}) is Boltzmann's constant, T (K) is the gas temperature and p (Pa) is the partial pressure of the vapor species i. If the cluster reaches a critical diameter, for which the addition of new molecules is larger than the removal, it becomes stable and will grow further giving origin to the new solid phase particle [74–77]. The critical cluster diameter d_p^{crit} (m) can be defined as

$$d_p^{crit} = \frac{4\sigma V_i}{k_B T \ln S} \tag{11}$$

and the rate of particle formation S_{nucl} ($m^3 s^{-1}$) is given by

$$S_{nucl} = \left(\frac{2\sigma}{\pi m_i}\right)^{1/2} \frac{V_i N_i^2}{S}\left(-\frac{16\pi V_i^2 \sigma^3}{3(k_B T)^3 (\ln S)^2}\right) \tag{12}$$

where m_i (kg) is the molecular mass of species i and N_i (m^{-3}) is the concentration of monomers. Typical alkali particle precursors found in small-scale pellets burners – K and Na chlorine and sulphate vapors – have a characteristically low saturation vapor pressure even at high temperatures. This means that supersaturation ($S > 1.0$) and critical cluster diameter can be easily reach at typical combustion temperatures. As discussed in the previous section, thermodynamic equilibrium calculations suggest that alkaline sulphates will nucleate before chlorines during the cooling of combustion gases (~950°C vs. ~600°C), when the conditions for nucleation are fulfilled [9,28,54]. These primary particles will act as seeds for the precipitation of condensable species [59].

4.4.2 Condensation

Heterogeneous condensation is a gas-to-particle process that affects total particle mass and size while total particle number is unchanged. Condensable species include K_2SO_4, KCl, $(KCl)_2$, K_2CO_3, Na_2SO_4, NaCl, $(NaCl)_2$, ZnO, $ZnCl_2$, PbO and $PbCl_2$ [71]. The molecular rate of condensation S_{cond} (m^3s^{-1}) of species i onto N_j particles of diameter size class j may be expressed as [76]:

$$S_{cond} = 2\pi D_i d_{p,j} N_j \frac{p_i - p_{i,s}}{R_{gas} T} K_{FS} \tag{13}$$

where D_i (m^2s^{-1}) is the diffusion coefficient, p_i (Pa) is the partial pressure of species i, $p_{i,S}$ (Pa) is the equilibrium vapor pressure over a droplet surface of diameter $d_{p,j}$ and K_{FS} is the Fruchs–Sutugin correction factor. Condensation is a particle growth processes that follows nucleation but condensation also occurs in the surface of soot particles.

4.4.3 Coagulation

Coagulation (agglomeration and aggregation) is a particle growing process affecting total particle number and size while the total particle mass remains constant. Particles move relative to each other and may collide and adhere to each other or coalesce. Coagulation may occur due to Brownian motion, leading to thermal coagulation, or external forces, leading to kinematic coagulation. Coagulation is governed by a general collision frequency function $\beta(v_m,v_n)$ (m^3s^{-1}) depending on the size of the colliding particles and on properties of the system such as temperature and pressure. Particles smaller than 1 μm, when the particle is larger than the mean free path of gas, will collide due to Brownian with the corresponding collision frequency function [74–77]:

$$\beta(V_m, V_n) = \frac{2k_B T}{3\mu}\left(\frac{1}{V_m^{1/3}} + \frac{1}{V_n^{1/3}}\right)\left(V_m^{1/3} + V_n^{1/3}\right) \tag{14}$$

where V_m (m^3) and V_n (m^3) are the particle volume of consecutive size classes m and n, respectively, and μ is the gas viscosity (Pa s). Larger particles, but much smaller than the mean free path of gas, will collide according to the collision frequency function derived from kinetic theory of gases for collisions among molecules that behave as rigid elastic spheres:

$$\beta(V_m, V_n) = \left(\frac{3}{4\pi}\right)^{1/6}\left(\frac{6k_B T}{\rho_p}\right)\left(\frac{1}{V_m} + \frac{1}{V_n}\right)^{1/2}\left(V_m^{1/3} + V_n^{1/3}\right) \tag{15}$$

Finally, the source term for coagulation for a specific size class k is as follows:

$$S_{coag,k} = 0.5 \sum_{m+n=k} \beta(V_m, V_n) N_m N_n - N_j \sum_{j=1}^{j} \beta(V_j, V_k) N_k \qquad (16)$$

where the first term of the RHS accounts for the generation rate due to smaller particles that will aggregate and populate size class k, and the second term accounts for the loss of particle in size class k that will aggregate and populate bigger size classes.

5 Particle Matter Emissions in Pellet-Fired Appliances

There has been a significant increase of studies on PM emissions from biomass pellets boilers and stoves in recent years, following earlier studies of gaseous emissions. It is worthwhile to mention that studies using laboratory reactors have contributed enormously to the better understanding of PM emission from biomass with relevance for pellet boilers and stoves. In this section, a short summary of the most relevant studies on PM emissions from pellet boilers/stoves and how they are influenced by operating conditions and fuel composition is shown. Table 5 shows a summary of relevant studies where measurements of PM emissions were made in pellet-fired boilers using different types of biomass. It is important to note that the appliances used in these studies come from many different manufacturers and may have significant design differences particularly in the combustion chamber and in the air supply. Furthermore, the operational set up of typical commercial boilers does not allow systematic variations of one parameter while keeping other constant; for instance, an increase in thermal load usually leads to a variation of the excess air [14,18]. The existence of several standards for measuring PM emissions and the absence of a common established protocol [78] have led to different ways of reporting emissions in terms of units (typically mg/MJ or mg/Nm^3) and normalization (some authors do not normalize, other may use different reference O_2 concentrations). It is not possible to establish quantitative correlations based on the data from different appliances; however, these data allow to establish qualitative correlations and to retrieve global trends giving insights into the processes that dominate PM emissions in small-scale commercial appliances.

The fuels used have covered a broad spectrum of biomass but with predominance of wood. Agricultural and forestry wastes from several sources have been used, coming from food production, food processing and wood industries with the goal of evaluating the potential of energetic valorization of the production waste [12,18,20,21]. The variety of the biomass used covers considerable ranges of ash content and ash composition, which allowed evaluating the effect of the inorganic content of the biomass on the emissions of PM [13,17–19]. All the boilers have a power lower than 50 kW and can be grouped into two categories: smaller boilers operated at a thermal load between a few kW and 20 kW [10,11,13–15,17–19] and bigger boilers operated at thermal loads between 30 and 50 kW [12,16,20,21].

Table 5: Summary of studies and measurements on PM emission from pellet-fired boilers and stoves.

Authors	Biomass	Feeding type	Operation mode	Thermal load	Excess air	Ash	Ash composition	Measurements
Johansson *et al.* [10]	Wood	Top	Continuous Modular	Variable 3–22 kW	n.a.	Constant 0.5%	n.a.	PM total (mg/MJ) PM MSD (mg/m^3 at 10% O_2) PM NSD (1/cm^3 at 10% O_2)
Bävfer *et al.* [11]	Wood	Top	Continuous	Variable 3–5 kW	Variable ratio 3.7–7.5	Constant 0.4%	Constant K: 351.6 mg/kg Na: 23.6 mg/kg	PM total (mg/MJ) PM 10 (mg/MJ) PM 2.5 (mg/MJ) PM MSD (mg/Nm^3) PM NSD (1/cm^3)
Verma *et al.* [12]	Apple Peat RCG SFH Straw	Top	n.a.	Variable 31.3–44.7 kW	Variable ratio 1.08–1.18	Variable 2.4%–10.7%	Variable S: 320–1390 mg/kg	PM total (mg/Nm^3)
Sippula *et al.* [13]	Pine bark/stem Birch bark/stem Spruce bark/stem Alder bark/stem Willow bark/stem	Top	Continuous	Constant 8 kW	Variable ratio 3.4–6.4	Variable 0.3%–4.0%	Variable K: 6.3–3875 mg/kg Na: 3.4–112 mg/kg S: 52.7–891 mg/kg Cl: 29.7–550 mg/kg	PM 10 (mg/MJ)
Fernandes and Costa [14]	Pine	Top	Continuous	Variable 10.8–21.3 kW	n.a.	Constant 1.3%	Constant K: 1550 mg/kg Na: 325 mg/kg Cl: 0.52 mg/kg	PM 10 (mg/Nm^3 at 13% O_2) PM 2.5 (mg/Nm^3 at 13% O_2) PM 1 (mg/Nm^3 at 13% O_2)

Bävfer *et al*. [15]	Oat	Top	Continuous	Constant 10 kW	n.a.	Constant 3.1%	Constant K: 5700 mg/kg Na: 100 mg/kg S: 1900 mg/kg Cl: 500 mg/kg	PM total (mg/MJ) PM MSD (mg/m^3 at 10% O_2) PM NSD ($\#/cm^3$ at 10% O_2) PM composition
Verma *et al*. [16]	Wood	Top Bottom	n.a.	Variable 15–32 kW	n.a.	Constant 0.21%	Constant S: 0.02 mg/kg	PM total (mg/Nm^3)
Boman *et al*. [17]	Sawdust Log residues Bark	Top Horizontal Bottom	Modular	Constant 3 kW	n.a.	Variable 0.61%–7.89%	Variable K: 600–3700 mg/kg Na: 100–1400 mg/kg S: 100–400 mg/kg Cl: 80–200 mg/kg	PM total (mg/MJ) PM MSD (mg/Nm^3) PM composition
Garcia-Maraver *et al*. [18]	Portuguese pine Spanish pine Cork Olive wood Olive Pruning	Top	Continuous	Variable ratio 14–17 kW	Variable ratio 5.8–12.2	Variable 0.8%–5.5%	Variable K: 920–4455 mg/kg Na: 206–10,780 mg/kg Cl: 3.2–550 mg/kg	PM 10 (mg/Nm^3 at 10% O_2) PM 2.5 (mg/Nm^3 at 10% O_2) PM 1 (mg/Nm^3 at 10% O_2) PM MSD (mg/Nm^3) PM composition

Table 5: Summary of studies and measurements on PM emission from pellet-fired boilers and stoves. (*Continued*)

Authors	Biomass	Feeding type	Operation mode	Thermal load	Excess air	Ash	Ash composition	Measurements
Tissari *et al.* [19]	Oat Wood Rape seed	Top	Continuous Modular	Variable 4.1–16.4 kW	n.a.	Variable 0.28%–4.22%	Variable K: 670–7910 mg/kg Na: 10.42–42.85 mg/kg S: 70–4350 mg/kg Cl: 41.8–578.39 mg/kg	PM 1 (mg/MJ) PM 10 (mg/MJ) PM NSD PM composition
Verma *et al.* [20]	Wood Apple Peat RCG CPW SFH	Top	n.a	Variable 29.8–44.7 kW	n.a.	Variable 0.65%–10.7%	Variable S: 105–1375 mg/kg	PM total (mg/Nm^3)
Verma *et al.* [21]	Wood Peat RCG CPW SFH Straw	Top	n.a.	Variable 11.21–35.5 kW	n.a.	Variable 0.21%–9.1%	Variable S: 0.02–1375 mg/kg	PM total (mg/Nm^3)

n.a., not available; MSD, mass size distribution; NSD, number size distribution; RCG, reed canary grass; CPW, citrus pectin waste; SFH, sunflower husk.

Although a considerable range of boilers with different powers is listed in Table 5, the great majority are top feed boilers and operated in a continuous mode. Some exceptions are a few studies using boilers with different types of feeding [16,17] and boiler operated at continuous or modular mode [10,17,19]. The excess air is quite an important parameter that relates to combustion conditions; however, it has not been common to measure it. PM concentrations were reported in mg/MJ and mg/Nm^3, with no clear preference of one over the other. Unfortunately, not all operating conditions and necessary parameters for inter-conversion were reported by all the authors; therefore, standardization of reported concentrations is not possible. The reader is referred to each study for the details of the experimental setup and measuring techniques.

The general finding amongst the listed studies is that PM emissions are strongly correlated with the ash content and ash composition [13,17–19], more than with operating conditions [18]. The PM 1 and PM 2.5 fractions have been found to be quite sensitive to the ash composition. The operating conditions do have an influence on PM emissions, particularly in the formation of soot and VOC when combustion conditions deteriorate, with the impact in the PM 1 and PM 2.5 fractions [14,18,22]. Disturbances in the fuel bed due to the type of feeding and modular operation bed have a significant impact in the PM 10 fraction due to the entrainment of char and ashes fragments in the flow field [10,16,19,61]. Nonetheless, the design of combustions chambers has a high level of maturity and combustion conditions are generally optimized. The efficient combustion of pellets results mainly in the emission of inorganic particles. In the following subsections, the effects of operating conditions and ash content and composition on PM emissions are summarized. Subsequently, the physical and chemical properties of fine particles are discussed. Finally, the methods for PM emissions reduction are discussed.

5.1.1 Influence of power, fuel feed and operation on PM emissions

Figure 8 shows the total PM emissions from pellet boilers with thermal loads ranging from 10 to 50 kW. Boilers tend to work in two ranges, 10–20 kW and 30–50 kW. The emissions vary between a few mg/Nm^3 and almost 700 mg/Nm^3. The data included in this plot were obtained for boiler operating in quite different operating conditions, operation mode and fuel type [46,48,52,61,62,73], thus giving an overview of how PM emission may vary. Most of the PM emissions are below 200 mg/Nm^3, which is still above the higher PM emissions thresholds established by European standards (see Table 1 and Section 2). It is important to note that boilers are optimized to operate with pellets that comply with strict quality standards and in most of the cited studies pellets that did not fully comply with these were used, particularly in terms of ash content. The use of biomass for pellets production from non-convectional sources, such as agricultural and forest wastes, is seen as desirable route; however, the properties of such biomass generally do not fulfil the established strict quality standards.

Figures 9 and 10 show the effect of operating conditions, namely thermal input, in the PM emissions. Note that in both the studies the same boiler was used and

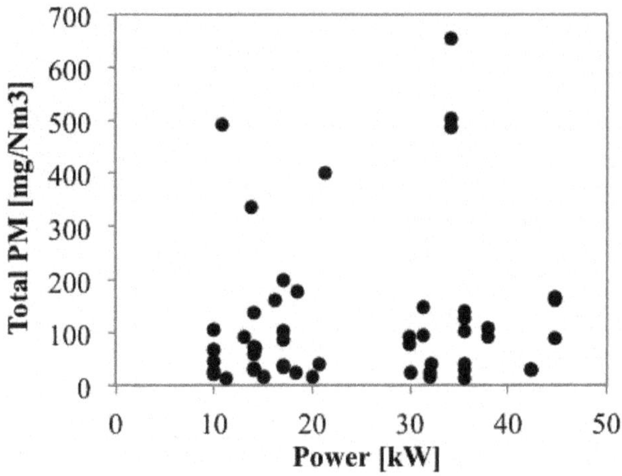

Figure 8: Total PM emissions from pellet boilers with thermal loads ranging from 10 to 50 kW [46,48,52,61,62,73].

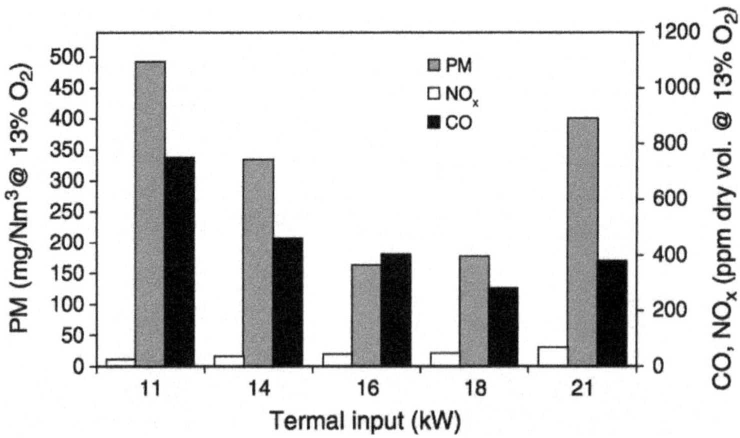

Figure 9: PM, NO_x and CO emissions as a function of the boiler thermal input [14].

that the increase in thermal input let to a decrease in the excess air and an increase in exhaust temperature. In Fig. 9, the effect of operating conditions on total PM emissions for pine pellets is evaluated. The PM emissions follow the same trend as CO emissions, suggesting that PM emissions are correlated with poor combustion conditions. Similar trends have been observed in the literature [7–9]. In Fig. 10, the effect of operating conditions and fuel type on PM1, 2.5 and 10 emissions is evaluated. The type of pellet has a more significant impact on the PM emissions than the thermal load. For all types of pellets, PM1 is dominant and PM2.5 is

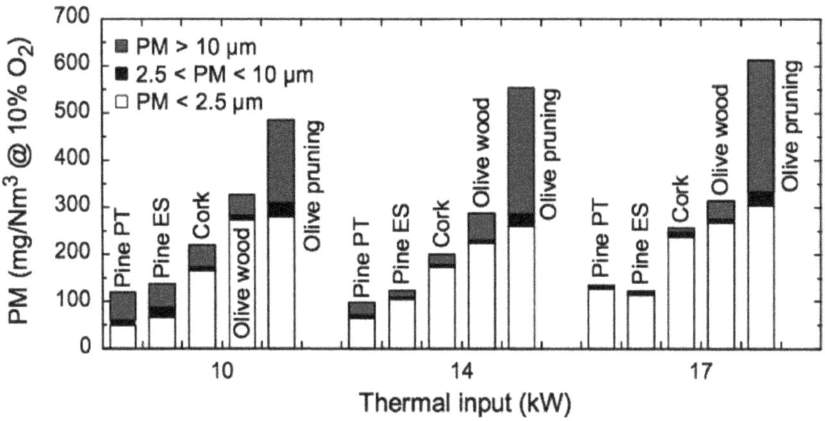

Figure 10: PM emissions of the different types of pellets for several conditions [18].

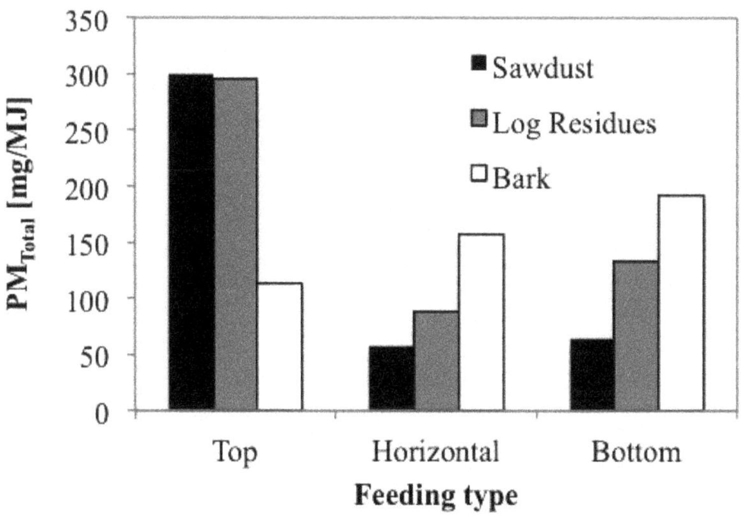

Figure 11: Effect of the feeding type on PM_{total} emissions in a small-scale pellets boiler using different biomass materials [17].

marginal. The PM1 emissions are significantly affected by the fuel ash composition as seen in the data for each thermal load. In the case of olive pruning pellets, the high amounts of PM10 are attributed to high content of ash that causes significant disturbances in the fuel bed. Figures 11 and 12 show the effect of the feeding type and the operating mode. It is clear that disturbance in the fuel bed led to an increase in PM emissions.

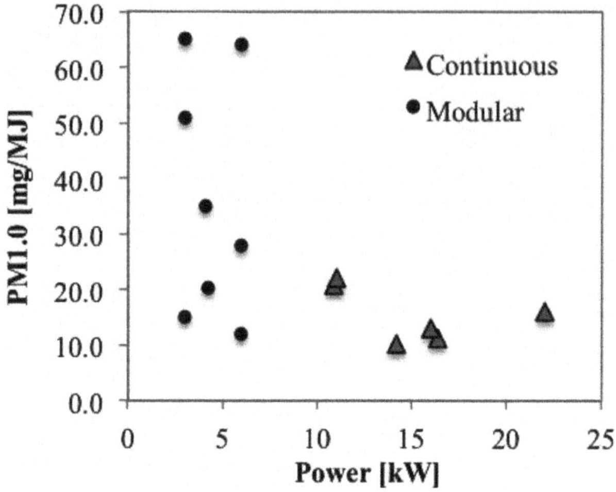

Figure 12: Effect of the operation mode on PM1.0 emissions in small-scale pellets boilers [10,19].

5.1.2 Influence of ash content and composition on PM emissions

The dominating elements found in the inorganic fraction of the fine fraction of PM emitted by pellet boilers are typically K, Na, S, Cl and to a smaller extent Zn and Fe [14,15,17–19]. These elements are found in the ash of the biomass and constitute their precursors; therefore, not surprisingly, the ash content and ash composition have a significant impact on PM emissions. Figure 13 shows the fuel alkali metal, chlorine and sulphur content obtained by Sippula *et al.* [13]. The latter found that the PM1.0 mass concentration correlated linearly with the ash content and composition, specifically the sum of the major PM forming elements. Furthermore, a linear correlation between PM1.0 mass concentration and each one of the elements aforementioned was observed. Other authors have also found a similar trend between PM emissions, including total emissions and other fractions, and the ash content and composition [12,15,17,18,20].

5.1.3 Physical and chemical properties of fine particles

The typical mass mean diameter of fine particles is consistently between 0.1 and 1 μm regardless of the type of boiler or the type of fuel [11,17–19] and the distribution is generally unimodal, with a few exceptions of bimodal distributions or even trimodal distributions within the range of 0.01 and 10 μm [18,19]. Figure 14 shows the mass size distributions obtained in two different studies where different types of curves can be observed for various fuels. The typical number mean diameter is closer to 0.1 μm [15,19], which means that although most of the mass of fine particles are concentrated in the range above 0.1 μm, most of the particles emitted are actually in the ultra-fine region. At different stages of the combustion cycle, the number and mass size distribution vary due to combustion conditions of fuel bed disturbances. Bävfer *et al.* [11] observed that PM emissions in terms

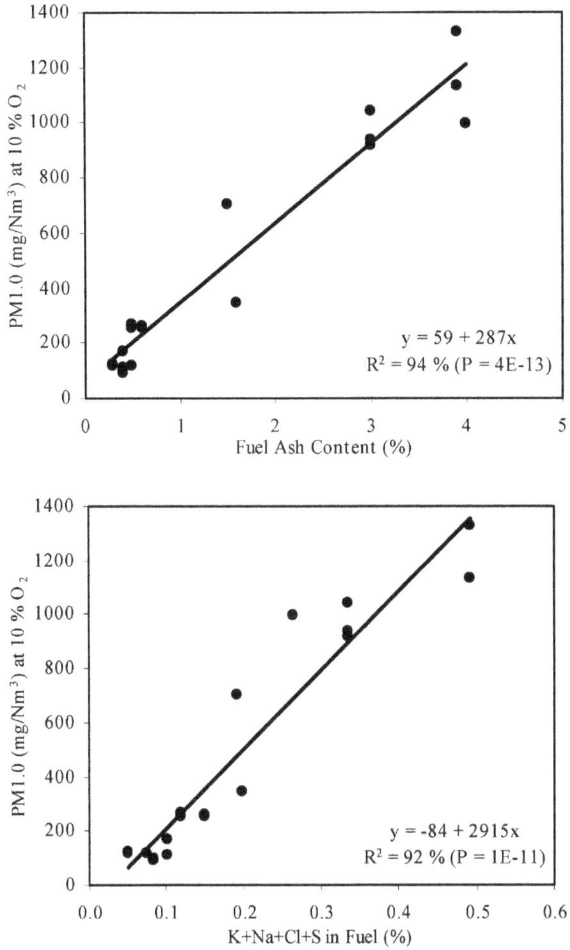

Figure 13: PM1.0 mass concentration as a function of ash content (top) and as a function of the fuel alkali metal, chlorine and sulphur content. Reprinted with permission from [13]. Copyright 2007 American Chemical Society.

of number and mass are higher during start-up. In the intermediate phase, we can observe a higher concentration of particles lower than 1 μm and a lower concentration of particles higher than 1 μm. The mean mass diameter also varied, with 0.4 μm for the start-up phase and 0.1 μm for the intermediate and the burnout phase. On the other hand, the shape of the number size distribution did not change.

The composition of the fine particles is once again consistent regardless of the type of boiler or the type of fuel [11,17–19]. Particles below 1 μm are dominated by K, S, Cl and Na, whereas particles above 1 μm may contain Mg, Ca, P, Fe and Si, which are components known to be retained in the char instead of being

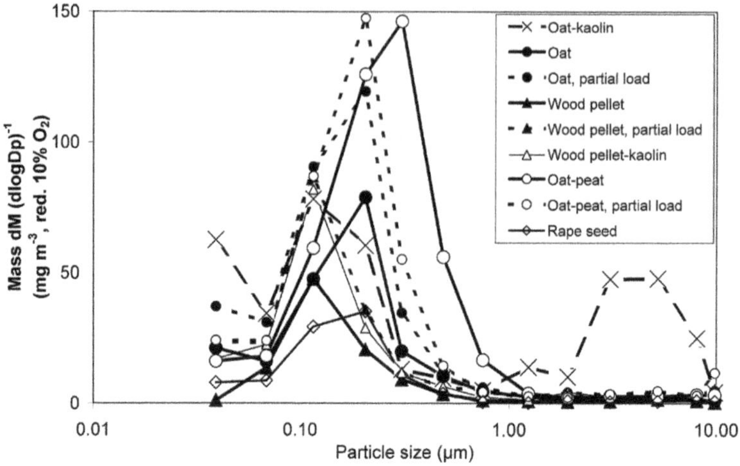

Figure 14: Mass size distribution of particles from the combustion of different wood and cereal-derived pellets. Reprinted with permission from [19]. Copyright 2008 American Chemical Society.

released to the gas phase. These results indicate that small fragments of char are entrained in the fluid flow and are emitted by the boiler.

5.1.4 Reduction of PM emissions

The reduction of PM emissions from boilers can be achieved via two ways: primary methods prevent the formation of particles during combustions, whereas in secondary methods particles are removed from the flue gas. The efficiency of primary reduction methods is associated with the origin and formation of particles, whereas the efficiency of secondary reductions methods is closely related with particle size and composition. There is no overall best method and a combination of methods may give best results.

Primary reduction measurements may be applied to organic or to inorganic particles. Organic particles comprise condensed hydrocarbons and soot and are a result of incomplete combustion; therefore, primary measures that optimize the combustion process will contribute to the reduction of PM emissions. To optimize the combustion process, it is necessary to find the optimal combination of turbulence, temperature and time [79] – the so-called 3T. Turbulence enhances mixing of fuel and oxidizer, which can be achieved by appropriate staging of primary and secondary air. High temperatures must be ensured as well as enough residence time allowing for high reaction rates high burnout. Computer fluid dynamics are an important tool for the development of advanced combustion chamber, and a few computational studies of domestic pellet boilers have emerged recently [80–82]. Porteiro *et al.* [80] obtained remarkable agreement between experiments and measurements, in particular flue gas temperature, boiler efficiency, and NO_x and CO emissions, showing the good potential of CFD to be used as a design tool for the optimization of combustion chambers. CFD may also be used for the study of inorganic particle formation, as it was done already for larger scale biomass combustion, and the development of primary measures at the combustion chamber – for instance, lowering the bed temperature to reduce inorganic vaporization. Another type of primary reduction method is the addition of sorbents to the biomass as a pre-treatment to inhibit the reaction paths that lead to the formation of inorganic particles [15,83]. Limestone, kaolin and calcite have been used and only kaolin has shown some potential in reducing the formation of fine particles. In both studies, the reduction is attributed to the binding of K to kaolinite, reducing its release to the gaseous phase.

Secondary measurements for PM emissions control include cyclone precipitators, catalytic converters, filters and electrostatic precipitators (ESP) [29,78,84]. Cyclone precipitators can only achieve the separation of particles above 5 μm with typical separation efficiency below 50% for particles below 10 μm [29]. Although not adequate for the reduction of fine PM emissions, cyclones can be used as first separation stage removing larger particles before a fine particle separation stage, therefore, reducing the particle load on the second stage. Catalytic converters can reduce the emission of organic particles by the oxidation of VOC and soot [78]. No promising results have shown so far as catalytic conversion requires high flue gas temperatures, meaning that it is not available during start-up when emissions are the highest [78,84]. The open structure of ceramic filters captures larger particles and the particles will experience a longer residence time leading to a higher burnout of organic particles [78]. The efficiency may vary between 10% and 70% (much depending on the filter geometry) and the filter is a low-cost technology of easy implementation. However, this technology does not have an impact on fine particles or VOC. Furthermore, the high-pressure drop caused by ceramic filters and catalytic converters has an acute negative effect on the combustion behavior of natural draft systems [84]. ESP is the most promising technology with efficiencies between 50% and more than 95%, since it is not fully established in small-scale pellet boilers performance evaluation over long periods is necessary [29,78,84]. Migliavacca *et al.* [85] evaluated the effect of ageing in two ESPs without cleaning systems by measuring PM emissions at the 1st, 5th and 10th day of a 10-day ageing test. The results revealed a decrease in separation efficiency from 90% to 20% due to the accumulation of PM matter and consequent reduction of active surface. Bologa *et al.* [86] tested an ESP in different appliances using different types of fuels. The separation efficiencies ranged from 73% to 88% with typical reductions of one order of magnitude of the total PM concentration. The ESP used had an automatic cleaning system and no 'ageing' affects were reported.

References

[1] Nussbaumer, T., Klippel, N. & Johansson, L., Survey on measurements and emission factors on particulate matter from biomass combustion in IEA countries, *Proceedigs of the 16th European Biomass Conference and Exhibition*, Valencia, pp. 2–6, 2008.

[2] Peksa, M., Dolzan, P., Grassi, A., Heinimö, J., Junginger, H.M., Ranta, T.M. & Walter, A., *Global Wood Pellets Markets and Industry: Policy Drivers, Market Status and Raw Material Potential*, IEA Bioenergy Task 40, 2007.

[3] Obaidullah, M., Bram, S., Verma, V. & De Ruyck, J., A review on particle emissions from small scale biomass combustion. *International Journal of Renewable Energy Research*, **2(1)**, pp. 147–159, 2012.

[4] Naeher, L.P., Brauer, M., Lipsett, M., Zelikoff, J.T., Simpson, C.D., Koenig, J.Q. & Smith, K.R., Woodsmoke health effects: a review. *Inhalation Toxicology*, **19(1)**, pp. 67–106, 2007.

[5] Buseck, P.R. & Pósfai, M., Airborne minerals and related aerosol particles: Effects on climate and the environment. *Proceedings of the National Academy of Sciences*, **96(7)**, pp. 3372–3379, 1999.

[6] Wiinikka, H., *High Temperature Aerosol Formation and Emission Minimisation During Combustion of Wood Pellets*. Doctoral Dissertation, Luleå University of Technology, 2005.

[7] Boman, C., *Particulate and Gaseous Emissions From Residential Biomass Combustion*. Doctoral Dissertation, Umeå University, 2005.

[8] Tissari, J., *Fine Particle Emissions from Residential Wood Combustion*. Doctoral Dissertation, University of Kuopio, 2008.

[9] Sippula, O., *Fine Particle Formation and Emission in Biomass Combustion*. Doctoral Dissertation, University of Kuopio, 2010.

[10] Johansson, L.S., Leckner, B., Gustavsson, L., Cooper, D., Tullin, C. & Potter, A., Emission characteristics of modern and old-type residential boilers fired with wood logs and wood pellets. *Atmospheric Environment*, **38(25)**, pp. 4183–4195, 2004.

[11] Bäfver, L.S., Leckner, B., Tullin, C. & Bernysen, M., Particle emissions from pellets stoves and modern and old-type wood stoves. *Biomass and Bioenergy*, **35(8)**, pp. 3648–3655, 2011.

[12] Verma, V.K., Bram, S., Delattin, F., Laha, P., Vandendael, I., Hubin, A., & De Ruyck, J., Agro-pellets for domestic heating boilers: Standard laboratory and real life performance. *Applied Energy*, **90(1)**, pp. 17–23, 2012.

[13] Sippula, O., Hytönen, K., Tissari, J., Raunemma, T. & Jokiniemi, J.K., Effect of wood fuel on the emissions from a top-feed pellet stove. *Energy & Fuels*, **21(2)**, pp. 1151–1160, 2007.

[14] Fernandes, U. & Costa, M., Particle emissions from a domestic pellets-fired boiler. *Fuel*, **103**, pp. 51–56, 2012.

[15] Bäfver, L.S., Rünnbäck, M., Leckner, B., Claesson, F. & Tullin, C., Particle emission from combustion of oat grain and its potential reduction by addition of limestone or kaolin. *Fuel Processing Technology*, **90(3)**, pp. 353–359, 2009.

[16] Verma, V.K., Bram, S., Dellatin, F. & De Ruyck, J., Real life performance of domestic pellet boiler technologies as a function of operational loads: a case study of Belgium. *Applied Energy*, **101**, pp. 357–362, 2013.

[17] Boman, C., Nordin, A., Böstrom, D. & Öhman, M., Characterization of inorganic particulate matter from residential combustion of pelletized biomass fuels. *Energy & Fuels*, **18(2)**, pp. 338–348, 2004.

[18] Garcia-Maraver, A., Zamorano, M., Fernandes, U., Rabaçal, M. & Costa, M., Relationship between fuel quality and gaseous and particulate matter emissions in a domestic pellet-fired boiler. *Fuel*, **119(1)**, pp. 141–152, 2014.
[19] Tissari, J., Sippula, O., Kuoki, J., Vuorio, K. & Jokiniemi, J.K., Fine particle and gas emissions from the combustion of agricultural fuels fired in a 20 kW burner. *Energy & Fuels*, **22(3)**, pp. 2033–2042, 2008.
[20] Verma, V.K., Bram, S., Gauthier, G. & De Ruyck, J., Evaluation of the performance of a multi-fuel domestic boiler with respect to the existing European standard and quality labels: Part-1. *Biomass and Bioenergy*, **35(1)**, pp. 80–89, 2011.
[21] Verma, V.K., Bram, S., Gauthier, G. & De Ruyck, J., Performance of a domestic pellet boiler as a function of operational loads: Part-2. *Biomass and Bioenergy*, **35(1)**, pp. 272–279, 2011.
[22] Fernandes, U. & Costa, M., Formation of fine particulate matter in a domestic pellet-fired boiler. *Energy & Fuels*, **27(2)**, pp. 1081–1092, 2013.
[23] Wiinikka, H. & Gebart, R., The influence of air distribution rate on particle emissions in fixed bed combustion of biomass. *Combustion Science Technology*, **117(9)**, pp. 1747–1766, 2005.
[24] Wiinikka, H. & Gebart, R., Critical parameters for particle emissions in small-scale fixed-bed combustion of wood pellets. *Energy & Fuels*, **18(4)**, pp. 897–907, 2004.
[25] Wiinikka, H. & Gebart, R., The influence of fuel type on particle emissions in combustion of biomass pellets. *Combustustion Science Technology*, **117(4)**, pp. 741–763, 2005.
[26] Wiinikka, H. & Gebart, R., Experimental investigations of the influence from different operating conditions on the particle emissions from a small-scale pellets combustor. *Biomass and Bioenergy*, **27(6)**, pp. 645–652, 2004.
[27] Wiinikka, H., Gebart, R., Boman, C., Boström, D., Nordin, A. & Öhman, M., High-temperature aerosol formation in wood pellets flames: spatially resolved measurements. *Combustion and Flame*, **147(4)**, pp. 278–293, 2006.
[28] Wiinikka, H., Gebart, R., Boman, C., Boström, D. & Öhman, M., Influence of fuel ash composition on high temperature aerosol formation in fixed bed combustion of woody biomass pellets. *Fuel*, **86(1)**, pp. 181–193, 2007.
[29] Nussbaumer, T., Overview on technologies for biomass combustion and emission levels of particulate matter, Prepared for Swiss Federal Office for the Environment (FOEN) as a contribution to the Expert Group on Techno-Economic Issues (EGTEI) under the Convention on Long-Range Transboundary Air Pollution (CLRTAP), Zürich, 2010.
[30] Nussbaumer, T., Combustion and co-combustion of biomass: fundamentals, technologies, and primary measures for emission reduction. *Energy & Fuels*, **17(6)**, pp. 1510–1521, 2003.
[31] Glasius, M., Ketzel, M., Wåhlin, P., Jensen, B., Mønster, J., Berkowicz, R. & Palmgren, F., Impact of wood combustion on particle levels in a residential area in Denmark. *Atmospheric Environment*, **40(37)**, pp. 7115–7124, 2006.

[32] Caseiro, A., Bauer, H., Schmidl, C., Pio, C.A. & Puxbaum, H., Wood burning impact on PM10 in three Austrian regions. *Atmospheric Environment*, **43(13)**, pp. 2186–2195, 2009.

[33] Szidat, S., Ruff, M., Perron, N., Wacker, L., Synal, H.-A., Hallquist, M., Shannigrahi, A.S., Yttri, K.E., Dye, C. & Simpson, D., Fossil and non-fossil sources of organic carbon (OC) and elemental carbon (EC) in Göteborg, Sweden. *Atmospheric Chemistry and Physics*, **9(5)**, pp. 1521–1535, 2009.

[34] Bari, M.A., Baumbach, G., Kuch, B. & Scheffknecht, G., Temporal variation and impact of wood smoke pollution on a residential area in southern Germany. *Atmospheric Environment*, **44(31)**, pp. 3823–3832, 2010.

[35] Johansson, L.S., Tullin, C., Leckner, B. & Sjövall, P., Particle emissions from biomass combustion in small combustors. *Biomass and Bioenergy*, **25(4)**, pp. 435–446, 2003.

[36] Carvalho, L., Wopienka, E., Pointner, C., Lundgren, J., Verma, V.K., Haslinger, W. & Schmidl, C., Performance of a pellet boiler fired with agricultural fuels. *Applied Energy*, **104**, pp. 286–296, 2013.

[37] Dockery, D.W., Pope III, C.A., Xu, X., Spengler, J.D., Ware, J.H., Fay, M.E., Ferris Jr., B.G. & Speizer, F.E., An association between air pollution and mortality in six U.S. cities. *New England Journal of Medicine*, **329(24)**, pp. 1753–1759, 1993.

[38] EN 303-05, Heating boilers Heating boilers for solid fuels, manually and automatically stoked, nominal heat output of up to 500 kW. Terminology, requirements, testing and marking, 2012.

[39] Verma, V.K., Bram, S. & De Ruyck, J., Small scale biomass heating systems: standards, quality labelling and market driving factors – an EU outlook. *Biomass and Bioenergy*, **33(10)**, pp. 1393–1402, 2009.

[40] Fiedler, F., The state of the art of small-scale pellet-based heating systems and relevant regulations in Sweden, Austria and Germany. *Renewable and Sustainable Energy Reviews*, **8(3)**, pp. 201–221, 2004.

[41] Johansson, L., Person, H. & Gustavsson, L., Measurement of particles from residential combustion of solid fuels-Nordic basis for a new coming CEN standard. Preliminary final report to the Nordic council of ministers, Working Group Sea and Air, 2006.

[42] Patashnick, H., Meyer, M. & Rogers, B., Tapered element oscillating microbalance technology. *Mine ventilation: Proceedings of the North American/ Ninth US Mine Ventilation Symposium*, Kingston, Ontario, pp. 625–631, 2002.

[43] SS028426(1991-12-04), Air quality, stationary source emissions – determination of concentration and mass flow rate of particulate material in gas-carrying ducts, Manual gravimetric method, 1991.

[44] SFS3866(1990-12-11), Air quality, stationary source emissions – determination of particulate emissions, manual method, 1990.

[45] EN 3058-2, Enclosed wood heaters smoke emission. Part 2: determination of particulate emission, 1994.

[46] PD 6434:1969, Recommendations for the design and testing of smoke reducing solid fuel burning domestic appliances, 1969.
[47] VDI 2066-Blatt2, Measurement of particulate matter, manual dust measurement in flowing gases, gravimetric determination of dust load, tubular filter devices (4 m^3/h, 12 m^3/h), 1993.
[48] VDI 2066-Blatt-5, Particulate matter measurement, dust measurement in flowing gases, particle size selective measurement by impaction method – cascade impactor, 1994.
[49] EN 13284-1, Stationary source emission – determination of low range mass concentration of dust – Part 1: manual gravimetric method, 2002.
[50] MEL02, Måaling af emissioner til luften, Bestemmelse af koncentrationen af total partikulært materiale i strøommende gas, Danish Environmental Agency Guideline, 2002.
[51] ISO 9096, Stationary source emissions – determination of concentration and mass flow rate of particulate material in gas-carrying ducts – manual gravimetric method, 1992.
[52] ISO 10155, Stationary source emissions – automated monitoring of mass concentrations of particles – performance characteristics, test methods and specifications, 1995.
[53] ASTME 2515, Standard test method for determination of particulate matter emissions collected by a dilution tunnel, 2011.
[54] Jiménez, S. & Ballaster, J., Influence of operating conditions and the role of sulfur in the formation of aerosols from biomass combustion. *Combustion and Flame*, **140(4)**, pp. 346–358, 2005.
[55] Nussbaumer, T., Czasch, C., Klippel, N., Johansson, L. & Tullin, C., *Particulate Emissions from Biomass Combustion in IEA Countries,* IEA Bioenergy Task 32, 2008.
[56] Pagels, J., Johansson, L., Hagström, M., Szpila, A., Bohgard, M., Tullin, C. & Sanati, M., Intercomparison of SMPS, ELPI and APS 3320 during sampling of particles emitted from a domestic wood pellet burner, *Sixth International Aerosol Conference*, Chinese Association for Aerosol Research in Taiwan, pp. 151–152, 2002.
[57] Boström, D., Skoglund, N., Grimm, A., Boman, C. & Boström, M., Ash transformation chemistry during combustion of biomass. *Energy & Fuels*, **26(1)**, pp. 85–93, 2011.
[58] CEN/TS 15290:2006, Solid biofuels – determination of major elements, 2006.
[59] Jöller, M., Brunner, T. & Obernberger, I., Modeling of aerosol formation during biomass combustion in grate furnaces and comparison with measurements. *Energy & Fuels*, **19(1)**, pp. 311–323, 2005.
[60] Gao, X. & Wu, H., Combustion of volatiles produced in situ from the fast pyrolysis of woody biomass: direct evidence on its substantial contribution to submicrometer particle (PM1) emission. *Energy & Fuels*, **25(9)**, pp. 4172–4181, 2011.

[61] Öhman, M., Boman, C., Hedman, H., Nordin, A. & Böstrom, D., Slagging tendencies of wood pellet ash during combustion in residential pellet burners. *Biomass and Bioenergy*, **27(6)**, pp. 585–596, 2004.

[62] Brunner, T., Bärnthaler, G. & Obernberger, I., Evaluation of parameters determining PM emissions and their chemical composition in modern residential biomass heating appliances. *Proceedings of the International Conference World Bioenergy*, pp. 81–86, 2008.

[63] Olsson, J.G., Jäglid, U., Pettersson, J.B. & Hald, P., Alkali metal emission during pyrolysis of biomass. *Energy & Fuels*, **11(4)**, pp. 779–784, 1997.

[64] Davidsson, K.O., Korsgren, J.G., Pettersson, J.B.C. & Jäglid, U., The effects of fuel washing techniques on alkali release from biomass. *Fuel*, **81(2)**, pp. 137–142, 2002.

[65] Knudsen, J.N., Jensen, A.P. & Dam-Johansen, K., Transformation and release to the gas phase of Cl, K, and S during combustion of annual biomass. *Energy & Fuels*, **18(5)**, pp. 1385–1399, 2004.

[66] Sippula, O., Hokkinen, J., Puustinen, H., Yli-Pirilä, P. & Jokiniemi, J.K., Particle emissions from small wood-fired district heating units. *Energy & Fuels*, **23(6)**, pp. 2974–2982, 2009.

[67] Westberg, H.M., Byström, M. & Leckner, B., Distribution of potassium, chlorine, and sulfur between solid and vapor phases during combustion of wood chips and coal. *Energy & Fuels*, **17(1)**, pp. 18–28, 2003.

[68] Kaufmann, H., *Chlorine-Compounds in Emissions and Residues from the Combustion of Herbaceous Biomass*. Doctoral Dissertation, Swiss Federal Institute of Technology, 1997.

[69] Sippula, O., Hokkinen, J., Lamberg, H., Puustinen, H., Yli-Pirilä, P., Tissari, J. & Jokiniemi, J.K., Particle emissions from small biomass and fuel oil fired heating units. *Proceedings of the 17th European Biomass Conference and Exhibition*, 2009.

[70] Obernberger, I., Brunner, T. & Jöller, M., Characterization and formation of aerosols and fly ashes from fixed bed biomass combustion. *Proceedings of the International Seminar on Aerosol from Biomass Combustion*, pp. 69–74, 2001.

[71] Jöller, M., Brunner, T. & Obernberger, I., Modeling of aerosol formation during biomass combustion for various furnace and boiler types. *Fuel Processing Technology*, **88(11)**, pp. 1136–1147, 2007.

[72] Lind, T., Kauppinen, E.I., Hokkinen, J., Jokiniemi, J.K., Orjala, M., Aurela, M. & Hillamo, R., Effect of chlorine and sulfur on fine particle formation in pilot-scale CFBC of biomass. *Energy & Fuels*, **20(1)**, pp. 61–68, 2006.

[73] Christensen, K.A., *The Formation of Submicron Particles from the Combustion of Straw*. Doctoral Dissertation, Technical University of Denmark, 1995.

[74] Agranosvski, I., *Aerosols: Science and Tecnology*. Wiley, 2011.

[75] Friedlander, S.K., *Smoke, Dust and Haze: Fundamentals of Aerosol Behavior*. Wiley: New York, 1977.

[76] Fuchs, N.A. & Sutugin, A.G., High-dispersed aerosols. In *Topics in Current Aerosol Research*, eds. G.M. Hidy and J.R. Brock, Pergamon Press: New York, pp. 1–60, 1971.

[77] Seinfield, J. & Pandis, S., *Atmospheric Chemistry and Physics*. John Wiley and Sons,1998.

[78] Mandl, C. & Obernberger, I., *Survey on the Present State of Particle Precipitation Devices for Residential Biomass Combustion with a Nominal Capacity up to 50 kW in IEA Bioenergy Task 32 member countries*, IEA Task 32, 2011.

[79] Brunner, T., Obernberger, I. & Biederman, F., Particulate matter emissions from small-scale biomass combustion systems – characterisation and primary measures for emission reduction, *IEA Bioenergy Conference*, Vienna, Austria, 2012.

[80] Porteiro, J., Collazo, J., Patin, D., Granada, E. & Moran, J.C., Numerical modeling of a biomass pellet domestic boiler. *Energy & Fuels*, **23**(**2**), pp. 2043–2051, 2009.

[81] Teixeira, J.C.F., Vasconcelos, B.N. & Ferreira, M.E.C., Simulation of a small scale pellet boiler. *ASME 2009 International Mechanical Engineering Congress and Exposition*, Lake Buena Vista, FL, USA, 2009.

[82] Chaney, J., Liu, H. & Li, J., An overview of CFD modelling of small-scale fixed-bed biomass pellet boilers with preliminary results from a simplified approach. *Energy Conversion and Managment*, **63**, pp. 149–156, 2012.

[83] Boman, C., Böstrom, D. & Öhman, M., Effect of fuel additive sorbents (kaolin and calcite) on aerosol particle emission and characteristics during combustion of pelletized woody biomass. *16th European Biomass Conference and Exhebition*, Valencia, pp. 1514–1517, 2008.

[84] Mandl, C., Obernberger, I. & Biederman, F., State-of-the-art and assessment of filter technologies for residential biomass combustion systems. *20th European Biomass Conference and Exhibition*, Milan, pp. 732–738, 2012.

[85] Migliavacca, G., Morreale, C., Hugony, F., Tombolato, I. & Pession G., Reduction of PM Emissions from Biomass Combustion Appliances: Evaluation of Efficiency of Electrostatic Precipitators. *Chemical Engineering Transactions*, **37**, pp. 25–30, 2014.

[86] Bologa, A., Paur, H., Ulbricht, T. & Woletz, K., Particle Emissions from Small Scale Wood Combustion Devices and their Control by Electrostatic Precipitation. *Chemical Engineering Transactions*, **22**, pp. 119–124, 2010.

CHAPTER 8

Slagging and Fouling Risks Derived from the Combustion of Solid Biofuels

Daniel J. Vega-Nieva[1], Angela Garcia-Maraver[2] & Luis Ortiz[1]
[1]*Forestry School, University of Vigo, Campus A Xunqueira, 36005 Pontevedra, Spain.*
[2]*Department of Civil Engineering, University of Granada, Campus de Fuente Nueva, 18071 Granada, Spain.*

Abstract

The interest in biomass combustion has grown exponentially in the last years, as it represents a means of renewable heat and energy that also promotes local development and mitigates climate change. However, the slagging and fouling risks of many potential feedstocks, given their high alkali, silica or chlorine contents, currently limit their application in combustion processes. This chapter presents various methods for monitoring biomass slagging and predicting fouling hazard. Discussion also addresses how the presence of such components may affect the formation of deposits in biomass boilers.

Keywords: Slagging, fouling, biomass combustion.

1 Introduction

One problem associated with biomass combustion concerns the ash generated in the process. Ash is a by-product that calls for appropriate management, and it moreover can lead to problematic phenomena such as slagging and fouling during the thermal conversion process itself [1]. It is, thus, a problem deriving both the quantity and the quality of the biomass ash. Both aspects are strongly influenced by the origin and characteristics of the biomass used [2,3].

Biomass composition, together with boiler design and operation, influences the composition of the ashes produced and consequently their final behaviour. Problems caused by low melting temperatures of the ashes may include fireside deposits of sintered or molten ash, which are due to the presence of very high contents of mineral oxides in some biomasses [4–7].

All facilities suffer economic loss from fireside deposits, but especially those firing more of the agricultural and urban fuels and less of the clean wood fuels. Although wood fuels are inherently low in slagging tendencies, adventitious materials such as clays and other soil components brought in with the fuel can also lead to fouling [8].

Ash deposits deteriorate burning, retard heat transfer, cause high-temperature corrosion and super-heater explosion, and can provoke mechanical failures [9–11], meaning greater maintenance costs and availability losses. Predicting sinter and slag formation in biomass combustion is essential to establish the biofuel quality and quality standards, and to determine which alternatives could minimize these effects in combustion furnaces [12].

2 Methods to Predict the Slagging and Fouling Tendencies of Biomass Fuels

Slagging occurs in the boiler sections that are directly exposed to flame irradiation. The mechanism of slag formation involves stickiness, ash melting and sintering. Slagging deposits consist of an inner powdery layer followed by silicate and alkali compounds [6,13].

On the other hand, fouling deposits are formed in the convective parts of the boiler. The mechanism of fouling is mainly due to the condensation of volatile species vaporized in previous boiler sections and is loosely bonded [13].

Slagging and fouling tendencies are directly related to the composition of the biomass fuels, as explained in this section.

2.1 Threshold values for critical elements of fuel composition and slagging and fouling indexes

2.1.1 Silicon

Fuels with high silicon content, either from intrinsic contamination or external contamination, generally show relatively high slagging tendencies [14–16]. In the study by Öhman *et al.* [15], a significant number of wood pellets reported to be either problematic or else problem-free regarding slagging were collected from ordinary residential pellet burners on the Swedish market. They showed that the problematic stem wood pellets had a significantly higher amount of Si in the fuel.

Si content in the fuel has correlated relatively well to the sintering tendencies in burners in several studies [14,17–19]. Exceptions to these general trends exist as well, i.e. fuels with very high Si/K ratios [16,17].

From these results, Öhman *et al.* [15] proposed a threshold of SiO_2 of 20%–25% of the fuel ash, the pellets with levels in or over this range being those that resulted in slagging problems in residential burners. Similarly, Vega-Nieva *et al.* [18–20] observed higher slagging tendencies in fuels with a lower basic ash percentage (e.g. higher amounts of silica and phosphorus in their composition); they proposed similar threshold values for the basic ash percentage in the fuel to prevent slagging. These indexes have been validated by combustion tests of agricultural and forest biomass fuels and have shown the potential to predict slagging occurrence and to design mixtures of fuels with a composition that minimizes slagging [19–21].

2.2 Alkali metals

Fuels with large contents of silica and potassium can result in the formation of K silicates [4,6,22], which create deposits in the boiler. Alkali metals react with silica contained in the ash, forming silicates with a very low melting point (<800°C) that deposit near the bed surfaces [23]. This is typical of some herbaceous fuels such as wheat or sorghum, fuels with potential slagging and fouling problems at temperatures above 700°C–800°C [24,25].

Miles *et al.* [26] found the alkali content of the fuel to be a key indicator of the slagging and fouling problems in the boilers and proposed the alkali index in eqn (1) to monitor the amount of alkali in fuels to minimize slagging and fouling occurrence

$$\text{AlkaliIndex} = (K_2O + Na_2O) \cdot Ash(\%) / HHV \quad (1)$$

here HHV is the higher heating value (MJ/Kg) at $H = 0\%$; and K_2O and N_2O are expressed as the percentage of these components in the fuel composition in d.b.

According to this index:

- Alkali index > 0.17 kg alkali/MJ $\rightarrow$ probable fouling.
- Alkali index > 0.34 kg alkali/MJ $\rightarrow$ fouling is certain to occur.

Nevertheless, some limitations have been identified when this index is used with low alkali fuels that present high melting risk at low temperatures, such as some agricultural shells [27]. Alternative alkali indexes have been presented by Fernandez Llorente *et al.* [12] or Vega-Nieva *et al.* [18–20,27,28], among others.

2.3 Chlorine

Chlorine plays a shuttle role, facilitating the transfer of potassium in the gas phase, where it participates in further reactions or is deposited on cold surfaces. On these surfaces, the chlorine is substituted by sulphur, forming potassium sulphate deposits and/or reacting with the metal's protective oxide layer, thereby initiating corrosion [13,23].

Bryers [6] and Pronobis [7], among others, have established the following values to limit Cl-based slagging and fouling in biomass fuels, where Cl represents the chlorine content (%) of the sample dry weight:

- $Cl < 0.2 \rightarrow$ low slagging trend.
- $0.2 < Cl < 0.3 \rightarrow$ medium slagging trend.
- $0.3 < Cl < 0.5 \rightarrow$ high slagging trend.
- $Cl > 0.5 \rightarrow$ very high slagging trend.

Obernberger *et al.* [29] gave a more conservative threshold, indicating that when Cl is higher than 0.1, there is a risk for corrosion and HCl emissions.

2.4 Ash melting temperatures

The ash fusibility test (AFT) is based on a continuous process that can be characterized by different temperatures according to norms ISO 540 [30] and DIN 51730 [31]. These temperatures are defined below and shown graphically in Fig. 1:

(a) *Deformation temperature (DT)*: Temperature at which the first signs of rounding occur, due to melting of either the tip or edges of the test piece.
(b) *Sphere temperature (ST)*: Temperature at which the edges of the test piece become completely round, the height becoming equal to the width of the baseline.
(c) *Hemisphere temperature (HT)*: Temperature at which the test piece is approximately hemispherical, the height being equal to half the base diameter.
(d) *Flow temperature (FT)*: Temperature at which the test piece material has spread out so that its height is one-third of that at the hemisphere temperature.

A large number of authors have proposed predictive functions of ash fusion temperatures from ash composition for coals [6]. Information on such predictive functions for biomass, however, remains scarce [32,33].

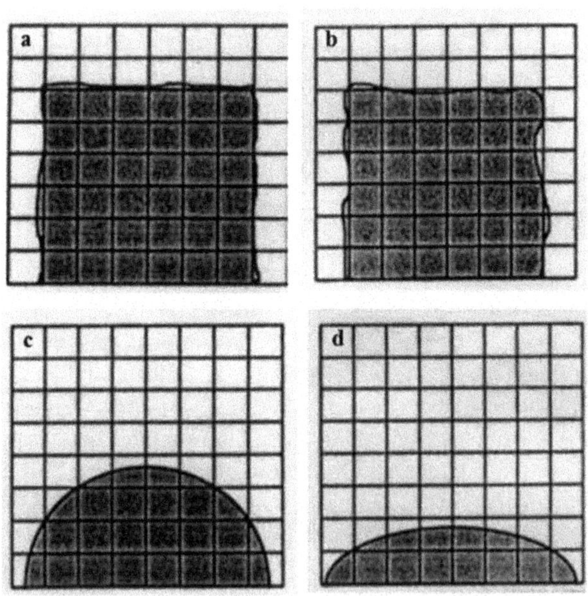

Figure 1: Ash fusibility temperatures [30,31]: (a) DT, (b) ST, (c) HT and (d) FT.

The standard AFT is extensively criticized in the literature, because of its lack of representativity, the low rate heating conditions utilized in the test, the low method repeatability and the lack of correspondence with observed boiler ash behaviour [34–39].

Hjuler [37] claims that the primary reason for unsuitability of the standard ash fusion test for biomass ashes is that biomass typically contains relatively high amounts of low melting components; these flow down in the test body due to gravity, leaving a skeleton of high melting minerals that may contain, for example, silicon and calcium. The initial low-temperature melt formation is not detected by the standard test because it does not affect the overall shape of the test body.

Many authors have, indeed, reported melting at temperatures below the standard measured ash fusion temperatures [34,38,39]. For instance, Skrifvars *et al.* [38] found that the ASTM standard ash fusion test generally gave temperatures 50°C–500°C higher than combustion tests. Gilbe *et al.* [39] found that the results from the standard ash fusion test [30] showed relatively high deformation temperatures in general, and therefore predicted a less problematic behaviour on the part of various biomass fuels in comparison to the actual slagging tendencies observed in controlled combustion experiments involving commercial pellet burner equipment.

2.5 Simple subjective slagging tests

Given the unreliability of ash fusion temperature tests for predicting slagging in biomass boilers during combustion processes, simple and fast slagging tests are used operationally by pellet and boiler manufacturers for the on-site quality control of fuels. Within the European project *Domoheat* – concerned with the combustion of Mediterranean biomass fuels in small and medium-sized pellet boilers – simple, rapid tests were conducted to study the combustion of a small amount of fuel at temperatures from 900°C to 1100°C, as shown in Fig. 2 [18,19,28].

Similar tests have been tested in the BIONORM and AshmelT projects [40,41]. As these methods can be useful for fast quality control of fuels in situ, they do not offer quantitative results and are subjected to a great degree of subjectivity in the interpretation of the results (e.g. YES/NO slagging scale).

2.6 Slag and ash granulometric (*SaG*) method

A quantitative slagging test based on the combustion of a representative amount of fuel was developed by the Faculty of Forest Engineering of the University of Vigo. The method entails the combustion of 200–250 g of biomass fuel in a 500-ml crucible following a defined temperature program reaching temperatures of 900°C–1100°C (Fig. 3). It offers a quantitative ranking of the slagging tendency of the fuels, based on the total calcinated weight percentage as well as the ash and slag size distribution. *SaG* method results were validated with combustion results from the *Domoheat* Project [18,19,28]. They are now undergoing further testing:

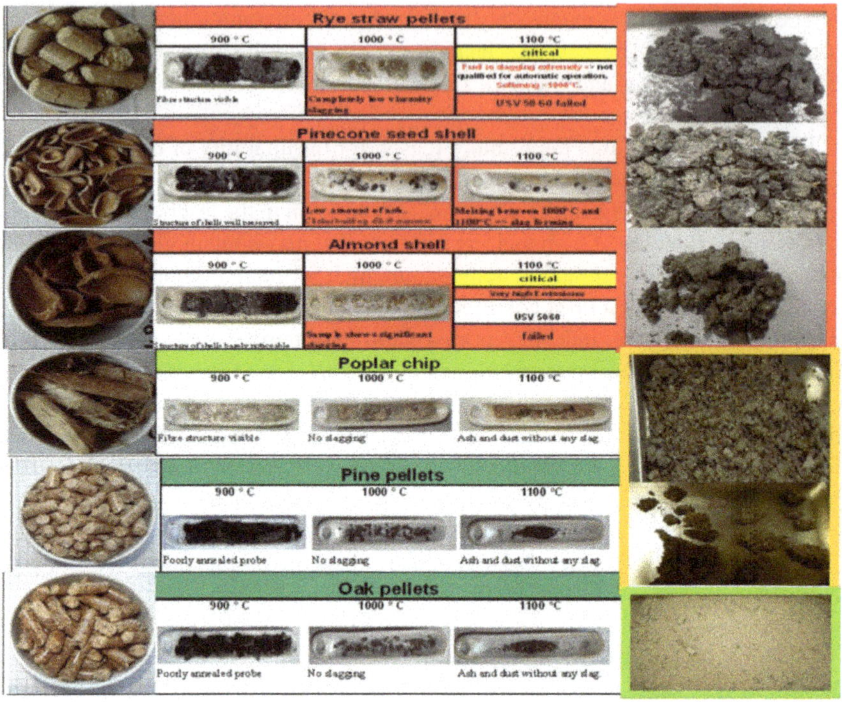

Figure 2: Simple rapid slagging test results for rye straw pellet, pinecone seed shell, almond shell, poplar chip, pine pellets and oak pellets from the *Domoheat* Project. Boiler bottom ash and slag are shown at the right picture. Slagging risk could be detected with simple rapid tests for rye straw, pinecone seed shell and almond shell samples (marked in red), which showed initial deformation temperatures <1,000°C. Neither ash fusion temperatures nor rapid tests could detect the moderate slagging present in soil-contaminated samples poplar chip and pine pellet (marked in green) [19].

the *AshmelT* project is an on-site operational test for measuring slagging risk of woody and herbaceous pellets requiring no special equipment [19,41,42].

This method has been utilized for the quality control of several wood chips and wood pellet SMEs within recent projects focusing on wood chip quality [21,43]. It has demonstrated great sensitivity for spotting moderate slagging tendencies caused by soil contamination and bark in woody biomass fuels (Fig. 4).

The *SaG* method has demonstrated utility to detect slagging in both woody and herbaceous fuels. These include herbaceous pellets of wheat straw, rapeseed or miscanthus, several types of shells such as almond shell, pinecone seed shell or olive stone, wood chips and wood pellets.

Figure 3: *SaG* method results from the *Domoheat* Project. Severe slagging was measured for rye straw, pinecone seed shell and almond shell fuels, with high amounts of calcinated weight and high percentage of melted and sintered fuel. The right figures show the granulometric distribution (>2, >1 and <1 mm) of the calcinated samples. Unlike ash fusion or rapid tests (Fig. 2), medium slagging tendency was measured with the *SaG* test for soil-contaminated fuels such as poplar chip and pine pellet, corresponding with observed slagging in the boiler combustion tests [18,19].

Currently subjected to a Round-robin test coordinated by *AshmelT* project partners at various European laboratories, the *SaG* test has been identified as a potential ISO standard for the measurement of slagging in the context of biomass fuels [42].

3 Conclusions

Biomass slagging represents one of the main barriers for the establishment of reliable certification systems and for the utilization of residual biomass and new

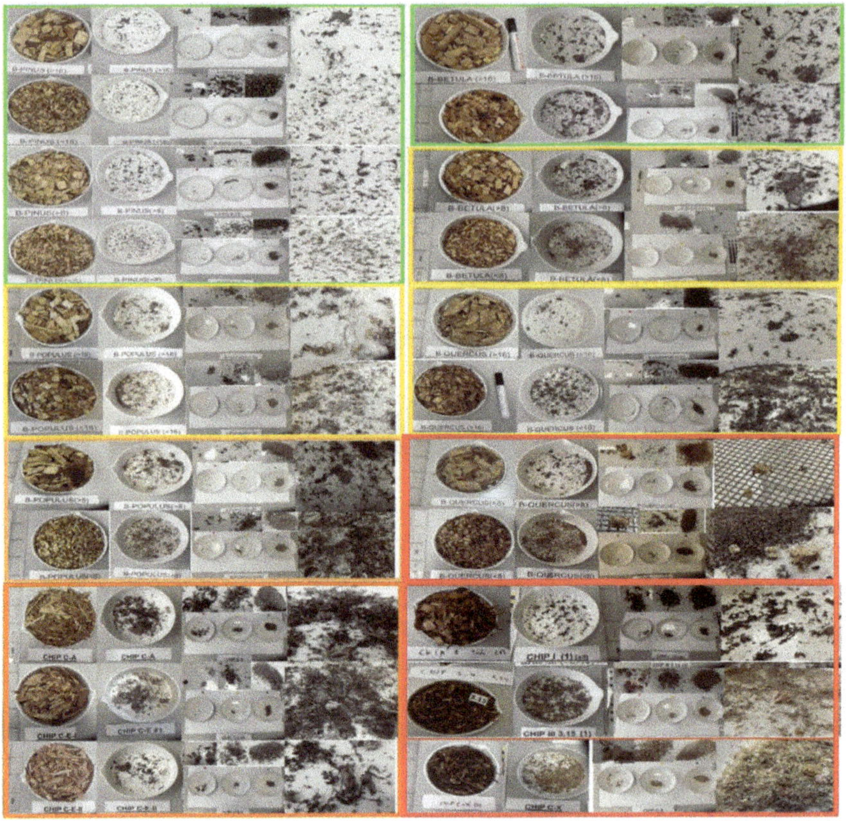

Figure 4: Results from the *SaG* test in the quality control of wood chips. The *SaG* test makes it possible to distinguish the formation of hard sinters from the presence of soil contamination (marked in red) and to quantify the presence of weak sinters caused by bark (marked in orange) in the wood chips from several biomass logistic centers in Spain [21,44].

energy crop biomass sources. Theoretical slagging indices based on ash composition have demonstrated the potential for monitoring and preventing biomass slagging, but further research is needed to validate such indices under varying combustion equipment technologies and varying operating conditions. A main barrier is the absence of a reliable ash slagging test to be included in the certification system of pellets, given the limitations of the current standard to provide reliable slagging risk estimations. New tests such as the *SaG* test have shown promise in the realm of biomass feedstock, including woody and herbaceous residual biomass, woody chips, agricultural shells and both woody and herbaceous pellets. Ongoing research on the validation of such methods may lead to the establishment of sound biomass slagging monitoring standards.

References

[1] Du, S., Yang, H., Qian, K., Wang, X. & Chen, H., Fusion and transformation properties of the inorganic components in biomass ash. *Fuel*, **117**, pp. 1281–1287, 2014.

[2] Van Loo, S. & Koppejan, J., *The Handbook of Biomass Combustion and Co-Firing*. Routledge: Sterling, VA, 2008.

[3] Tortosa Masiá, A.A., Buhre, B.J.P., Gupta, R.P. & Wall, T.F., Characterising ash of biomass and waste. *Impacts of Fuel Quality on Power Production*, **88**, pp. 1071–1081, 2007.

[4] Werther, J., Saenger, M., Hartge, E.-U., Ogada, T. & Siagi, Z., Combustion of agricultural residues. *Progress in Energy and Combustion Science*, **26**, pp. 1–27, 2000.

[5] Demirbas, A., Potential applications of renewable energy sources, biomass combustion problems in boiler power systems and combustion related environmental issues. *Progress in Energy and Combustion Science*, **31**, pp. 171–192, 2005.

[6] Bryers, R.W., Fireside slagging, fouling, and high-temperature corrosion of heat-transfer surface due to impurities in steam-raising fuels. *Progress in Energy and Combustion Science*, **22**, pp. 29–120, 1996.

[7] Pronobis, M., Evaluation of the influence of biomass co-combustion on boiler furnace slagging by means of fusibility correlations. *Biomass and Bioenergy*, **28**, pp. 375–383, 2005.

[8] Baxter, L.L., Miles, T.R., Jenkins, B.M., Milne, T., Dayton, D., Bryers, R.W., *et al.*, The behavior of inorganic material in biomass-fired power boilers: field and laboratory experiences. *Fuel Processing Technology*, **54**, pp. 47–78, 1998.

[9] Aho, M. & Silvennoinen, J., Preventing chlorine deposition on heat transfer surfaces with aluminium–silicon rich biomass residue and additive. *Fuel*, **83**, pp. 1299–1305, 2004.

[10] Szemmelveisz, K., Szűcs, I., Palotás, Á.B., Winkler, L. & Eddings, E.G., Examination of the combustion conditions of herbaceous biomass. *Fuel Processing Technology*, **90**, pp. 839–847, 2009.

[11] Knudsen, J.N., Jensen, P.A. & Dam-Johansen, K., Transformation and release to the gas phase of Cl, K, and S during combustion of annual biomass. *Energy & Fuels*, **18**, pp. 1385–1399, 2004.

[12] Fernández Llorente, M.J. & Carrasco García, J.E., Comparing methods for predicting the sintering of biomass ash in combustion. *Fuel*, **84**, pp. 1893–1900, 2005.

[13] Tortosa-Masiá, A.A., Ahnert, F., Spliethohh, H., Loux, C.J. & Hein, K.R.G. Slagging and fouling in biomass co-combustion. *Thermal Science*, **9**, pp. 85–98, 2005.

[14] Öhman, M., Boman, C., Hedman, H., Nordin, A. & Boström, D., Slagging tendencies of wood pellet ash during combustion in residential pellet burners. *Biomass and Bioenergy*, **27**, pp. 585–596, 2004.

[15] Öhman, M., Nordin, A., Hedman, H. & Jirjis, R., Reasons for slagging during stemwood pellet combustion and some measures for prevention. *Biomass and Bioenergy*, **27**, pp. 597–605, 2004.

[16] Öhman, M., Gilbe, C., Näzelius, I.L. Hedman, H., Boström, D., Boman, C., *et al.*, Slag formation during combustion of biomass fuels with low phosphorus content. *From Research to Industry Markets: Proceedings of the 19th European Biomass Conference and Exhibition*, Berlin, Germany, 6–10 June 2011, pp. 1267–1270, 2011.

[17] Gilbe, C., Öhman, M., Lindström, E., Boström, D., Backman, R., Samuelsson, R., *et al.*, Slagging characteristics during residential combustion of biomass pellets. *Energy & Fuels*, **22**, pp. 3536–3543, 2008.

[18] Vega-Nieva, D., Modeling of forest energy crops growth in Galicia, NW Spain and definition of biomass quality characteristics for combustion in small and medium size pellet boilers, Vigo, 2011.

[19] Vega-Nieva, D., Alvarez, C. & Ortiz, L., Results of new laboratory methods and slagging classification systems for the prediction and quantification of ash slagging in woody and herbaceous biomass fuels. *Central European Biomass Conference*, 2014.

[20] Vega-Nieva, D., Dopazo, R. & Ortiz, L., Strategies for minimizing ash slagging in combustion of mediterranean biomasses. *Proceedings of World Bioenergy*, 2012.

[21] Somoza, L., Vega-Nieva, D. & Ortiz, L., Quality control of wood chips and wood pellet from the biomass logistic center of Biopalas, 2014.

[22] Jenkins, B.M., Baxter, L.L., Miles, Jr. T.R. & Miles, T.R., Combustion properties of biomass. *Fuel Processing Technology*, **54**, pp. 17–46, 1998.

[23] Arvelakis, S., Vourliotis, P., Kakaras, E. & Koukios, E.G., Effect of leaching on the ash behavior of wheat straw and olive residue during fluidized bed combustion. *Biomass and Bioenergy*, **20**, pp. 459–470, 2001.

[24] Vassilev, S.V., Baxter, D. & Vassileva, C.G., An overview of the behaviour of biomass during combustion: Part II. Ash fusion and ash formation mechanisms of biomass types. *Fuel*, **117**, pp. 152–183, 2014.

[25] Vassilev, S.V., Baxter, D., Andersen, L.K. & Vassileva, C.G., An overview of the chemical composition of biomass. *Fuel*, **89**, pp. 913–933, 2010.

[26] Miles, T.R., Baxter, L.L., Bryers, R.W., Jenkins, B.M. & Oden, L.L., Boiler deposits from firing biomass fuels. *Biomass and Bioenergy*, **10**, pp. 125–138, 1996.

[27] Vega-Nieva, D., Dopazo, R. & Ortiz, L., Slagging risk of Mediterranean biomasses for combustion. *World Bioenergy*, 2010.

[28] Vega-Nieva, D., Dopazo, R. & Ortiz, L., Domoheat project: slagging risk of mediterranean biomasses in combustion in domestic pellet boilers. *Expobioenergia*, 2010.

[29] Obernberger, I., Brunner, T. & Barnthaler, G., Chemical properties of solid biofuels – significance and impact. *Biomass and Bioenergy*, **30**, pp. 973–982, 2006.

[30] ISO-540. Hard coal and coke: determination of ash fusibility, 2008.

[31] DIN51730. Determination of fusibility of fuel ash, 1998.
[32] Friedl, A., Padouvas, E., Rotter, H. & Varmuza, K., Prediction of heating value of biomass fuel and ash melting behaviour using elemental compositions of fuel and ash. *9th International Conference on Chemometrics in Analytical Chemistry*, 2004.
[33] Seggiani, M., Empirical correlations of the ash fusion temperatures and temperature of critical viscosity for coal and biomass ashes. *Fuel*, **78**, 1121–1125, 1999.
[34] Huggins, F.E., Kosmack, D.A. & Huffman, G.P., Correlation between ash-fusion temperatures and ternary equilibrium phase diagrams. *Fuel*, **60**, pp. 577–584, 1981.
[35] Huffman, G.P. & Huggings, F.E., Fouling and slagging resulting from impurities in combustion gases. *Proceedings of the 1981 Engineering Foundation Conference*, July 12–17, New England College, Henniker, New Hampshire, 1983.
[36] Van der Drift, A. & Olsen, A., Conversion of biomass: prediction and solution methods for ash agglomeration and related problems, 1999.
[37] Hjuler, K., Prediction of fouling and sintering in thermal conversion of biomass – determination of ash fingerprints. *1st World Conference on Biomass for Energy and Industry*, Sevilla, Spain, 5–9 June 2000, Vol. 1, p. 2135, 2001.
[38] Skrifvars, B.-J., Öhman, M., Nordin, A. & Hupa, M., Predicting bed agglomeration tendencies for biomass fuels fired in FBC boilers: a comparison of three different prediction methods. *Energy & Fuels*, **13**, 359–363, 1999.
[39] Gilbe, C., Lindström, E., Backman, R., Samuelsson, R., Burvall, J. & Öhman, M., Predicting slagging tendencies for biomass pellets fired in residential appliances: a comparison of different prediction methods. *Energy & Fuels*, **22**, pp. 3680–3686, 2008.
[40] Junker, H., Nikolaisen, L., Møller, H., Jensen, P.D., Hjuler, K., Dahl, J., *et al.*, Characterization of solid biofuels 2004 – development of methods, PSO project no. 5297, 2008.
[41] Schön, C., Hartmann, H., Schwabl, M.S.F., Dahl, J., Rathbauer, J., *et al.*, New evaluation strategies regarding slag prediction in pellet boilers. *World Bioenergy*, 2014.
[42] Rathbauer, J., Schön, C., Vega-Nieva, D., Hartmann, H. & Ortiz, L., Simplified method for slag prediction – PASSA method. *Pellets Conf. Berlin*, 2014.
[43] Gallego, A., Vega-Nieva, D.J. & Ortiz, L., Quality control of wood chips from the logistic biomass centres Foresga in Galicia NW Spain, 2014.
[44] Gallego, A., Quality control of wood chips from the logistic biomass centre of Foresga in Galicia NW Spain at two seasons of harvest, Vigo, 2014.

CHAPTER 9

The Use of Biomass in District Heating and Cooling Systems

Jose Antonio La Cal Herrera
Department of Management Enterprises, Marketing and Sociology, University of Jaen, Campus las Lagunillas, s/n, 23071 Jaen, Spain.

Abstract

Because of their environmental and economic advantages, district heating and cooling systems fed with biomass are of increasing interest worldwide. Their implementation not only reduces greenhouse gases due to use of a renewable energy resource but also reduces exploitation costs, benefitting from a centralized energy system. In addition, these systems entail significant advantages for the buildings connected to them in terms of safety, electricity demand, space availability and energy rating. Due to all of this, it is expected that one of the most relevant advancements in biomass use in the coming years will be its energy valorization by means of DHC systems.
Keywords: Biomass, district heating and cooling, efficiency, savings, profitability.

1 Introduction

The first district heating system was put into operation in Lockport, New York, in 1877. This system used a steam boiler and a distribution net of 4.8 km, which supplied heat to several industries and residential buildings. Nevertheless, the Romans were the real precursors of these systems: by means of wooden pipes, they distributed hot water among the rooms in buildings [1].

In 1983, the Hamburg council in Germany installed a centralized heating system that used the residual thermal energy of a central power station. Similar examples progressively occurred in the USA, Scandinavia and Denmark until the end of the Second World War, when the competitiveness with fossil fuels made the viability of the biomass systems disappear. This fact, together with the construction of large power stations in the outskirts of cities, notably increased the transport costs of thermal energy that in the majority of the cases was residual steam. This situation reverted again in 1970 because of the oil crisis; it was then that countries such as Sweden or Denmark started to wager on the incineration of municipal solid waste and the use of renewable energies, especially that from biomass.

District cooling systems are more recent. This technology was developed in the USA during the 1960s and its evolution has been provoked by the increment of the thermal energy demand in buildings from the tertiary sector (hospitals, faculties, etc.) and the increasing requirements of thermal comfort in the population.

These district heating and cooling (DHC) systems were also implemented in Asia beginning in the 1970s because of the high cooling demand in office buildings. Currently, 50% of cities in China have centralized systems, which are exponentially increasing because of high electricity prices and the existing energy deficiency.

An increasing and global evolution of these systems, which are instituted in almost every part of the world, is thus observed.

2 Legislative Framework

At the European level, the main norm in matters of centralized systems is the directive 2010/31/CE [2] on the energy performance of buildings. This directive defines the concept of centralized urban heating system; in article 6 it states that: 'Member States shall take the necessary measures to ensure that new buildings meet the minimum energy performance requirements set in accordance with Article 4. For new buildings, Member States shall ensure that, before construction starts, the technical, environmental and economic feasibility of high-efficiency alternative systems such as those listed below, if available, is considered and taken into account: decentralised energy supply systems based on energy from renewable sources; cogeneration; district or block heating or cooling, particularly where it is based entirely or partially on energy from renewable sources; heat pumps.'

Moreover, in article 20, it is noted that: 'Member States shall ensure that guidance and training are made available for those responsible for implementing this Directive. Such guidance and training shall address the importance of improving energy performance, and shall enable consideration of the optimal combination of improvements in energy efficiency, use of energy from renewable sources and use of district heating and cooling when planning, designing, building and renovating industrial or residential areas.'

Other legislative challenges from this directive that should be highlighted are summarized below:

- By the end of 2018, all new public buildings have to be 'buildings with almost null energy consumption,' and this requirement will be extended to all new private sector buildings by the end of 2020.
- Energy efficiency certificates are already mandatory for buildings that are going to be sold or rented.
- There is a positive influence of DHC systems on the energy performance of a building that has to be taken into account.

On the other hand, the new energy efficiency directive 2012/27/EU [3] cites that 'district heating and cooling has significant potential for saving primary energy, which is largely untapped in the Union' and introduces references to these systems such as their contributions to improving energy efficiency and the reduction of greenhouse gas emissions.

3 DHC Systems

3.1 Definition

According to the European directive 2010/31/CE [2], 'district heating or district cooling means the distribution of thermal energy in the form of steam, hot water or chilled liquids, from a central source of production through a network to multiple buildings or sites, for the use of space or process heating or cooling.'

To sum up, the DHC systems are pipe systems that connect the energy sources to the energy consumption points. Their main advantage is the efficient use of thermal energy for the acclimatization of buildings and/or facilities. In addition, they are capable of taking advantage of the residual heat from thermal and/or cogeneration plants, industrial processes and renewable energies. Driven by the ability to provide synergies between local resources and thermal sinks, DHC systems have interfaces not only with a huge variety of other energy and non-energy sectors (Fig. 1), and function therefore with close collaboration from other stakeholders, such as the renewable industry (solar, geothermal and biomass, including waste), building owners, operators and users, industrial facilities and the service sector, but also with urban planners and local authorities [4].

3.2 DHC systems fed with biomass

To achieve a competitive district system from an economic and sustainable point of view, it is necessary not only to use efficient technologies but also to use residual and/or renewable sources of energy such as biomass.

The only difference between a conventional system and a DHC system fed with biomass is the power plant, which is the main element of the system because its objective is to satisfy the thermal demand. This plant has to be dimensioned and designed according to the fuel and technology to be used and its location.

Regarding biomass fuels, the most commonly used is lignocellulosic biomass with moisture content lower than 20%. For economic reasons, biomass is generally used in chips (100–150 €/tonne), although pellets (300 €/tonne) are an option to consider when the heterogeneity of the biomass supply can alter the behaviour of the plant (Fig. 2) and, in consequence, the energy supply [5].

The technical requirements necessary to use biomass in the DHC systems are generally established by the manufacturers of the boilers and are focussed on the particle size and on the moisture and ash contents. Table 1 shows the characteristics of a biomass sample suitable for use in these centralized systems.

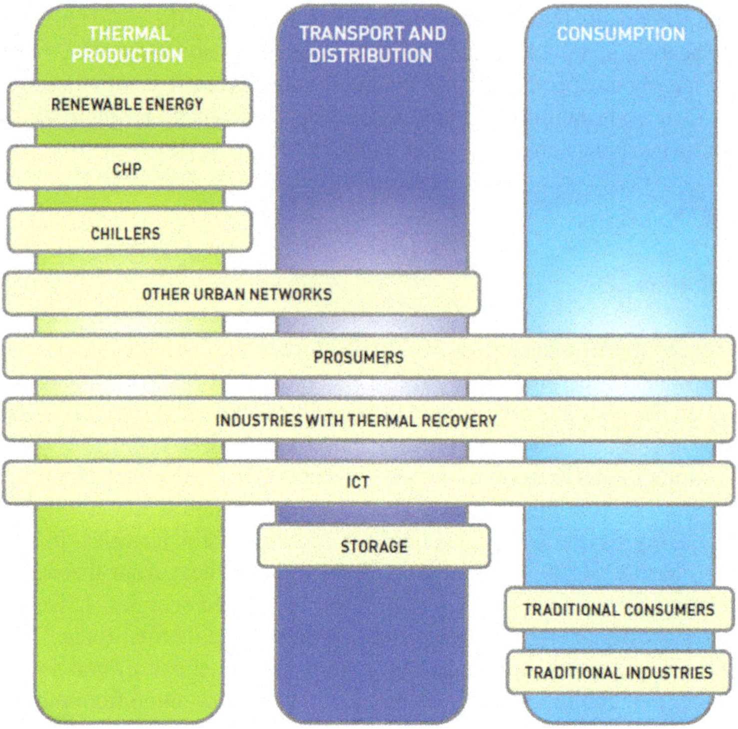

Figure 1: Horizontal nature of DHC systems [4].

Figure 2: Biomass chips and biomass pellets.

3.3 Advantages and disadvantages

DHC systems have several advantages in comparison with the individualized systems. Among them, the most relevant are listed below:

- Economic savings for the consumer due to the reduction of the necessary power and the final price of the energy consumed.
- Sustainability because of using renewable sources of energy.

Table 1: Characteristics of a biomass sample suitable for use in DHC systems according to the manufacturer [6].

Parameter	Suitability range
Size (mm)	0–5
Moisture content (%)	6–15
Lower heating value (kcal/kg)	3,800–4,250
Ash content (%)	0.5–2.5
Volatile content (%)	70–80
Sulphur content (%)	<0.15
Chlorine content (%)	<0.2
Fixed carbon (%)	<30

- Flexibility to be adapted to the energy demand.
- Efficiency related to the higher yield and the better insulation of the pipes.
- Higher value of the building because of its better energy rating.
- Easy management caused by the better relationship between promoters and clients.

These systems have some disadvantages such as:

- The necessity of an agreement between all the agents: administrations, promoters, users, enterprises, etc.
- The infrastructure necessary to connect the power plant to the final consumers.

4 The Model of Energy Service Companies

One of the drivers of the DHC systems is the so-called Energy Service Company (ESCO) model [7]. This model is based on the definition of 'a natural or legal person that delivers energy services and/or other energy efficiency improvement measures in a user's facility or premises, and accepts some degree of financial risk in so doing' established by the directive 2006/32/CE on energy end-use efficiency and energy services [8].

The payment for the services delivered is based (either wholly or in part) on the achievement of energy efficiency improvements and on the meeting of the other agreed performance criteria established in the energy performance contract (Fig. 3).

The key characteristics of this type of enterprise or ESCO are as follows [9]:

- The ESCO guarantees the energy savings and/or the provision of the same level of energy service at a lower cost by implementing energy efficiency measures.
- The remuneration of the ESCO is directly tied to the energy savings achieved.
- The ESCO can either finance or assist in arranging financing for the installation of an energy project they implement by providing a savings guarantee.

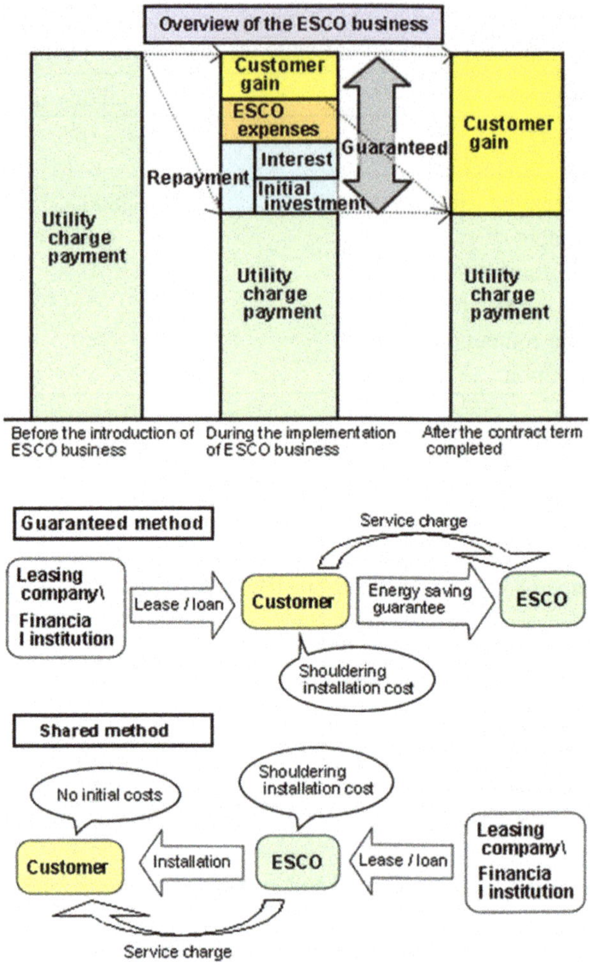

Figure 3: Overview of the ESCO business [10].

- The ESCO retains an on-going operational role in measuring and verifying the savings over the financing term.
- ESCOs offer a large range of energy services including: energy analysis and audits; energy management; project design and implementation; maintenance and operation; power generation and energy supply; monitoring and evaluation; facility and risk management.

ESCOs can be grouped into different ownership groups [9]:

- ESCOs that originate from equipment manufacturers and suppliers.
- ESCOs that stem from energy utilities or supply companies, public sector agencies or public–private joint ventures.
- Independent ESCOs.

5. The Case of Geolit, a DHC Central System Using Biomass

Geolit Acclimatization is a company created with the objectives of design, procurement and construction of an efficient and innovative DHC system, with benefits for its users and for the environment. This DHC system was the first one constructed in Spain. It is located in the province of Jaen, where there is a high energy potential from residual biomass from the olive oil sector. In accordance with this, the facility was designed for the use of this biomass as a fuel.

The thermal installations of Geolit work to supply hot and cold water for 35,000 m^2 of acclimatized buildings [11,12].

5.1 Specifications of the facility

5.1.1 Thermal production building

The thermal production building (Fig. 4) is an independent building where one can find boilers, chillers, pumps and cooling towers for the production and distribution of hot and cold water.

The installation includes two biomass boilers specifically designed for olive stone and olive chip combustion as well as forest and agricultural residues or energy crops.

5.1.2 Hot water production: biomass boilers

Hot water is produced through the operation of two biomass boilers (Fig. 5). The total installed thermal power is 6,000 kW, well above the initial thermal demands (2,500 kW estimated). Both boilers are located in the central heating room.

The biomass building was created and installed with a modular design and automatic operation, in which each of the modules has the characteristics described in Table 2.

The biomass boiler complex is associated with a huge biomass storage silo (450 m^3). The installations are controlled by an electronic control system located in a separated room.

Figure 4: Thermal production building at Geolit.

Figure 5: Biomass boilers (3,000 kW/boiler) at Geolit.

Table 2: Characteristics of the biomass boilers at Geolit.

Thermal power (MW)	3.0
Water flow (l/s)	103.2
Temperature operation range (°C)	105–80

5.1.3 Cold water production: single-effect absorption chillers

The hot water boiler outlet, at 90°C, is used like an energy inlet in the single-effect absorption chiller. These equipments use a LiBr-like absorbent and can provide 4,000 kW in the form of cold water at 5.5°C. Therefore, this equipment produces cold water from hot water without a significant use of electricity.

In addition to the 4,000 kW single-effect absorption chiller another absorption chiller (backup system) is available, which can provide 2,000 kW in the form of cold water at 5.5°C (Fig. 6).

The absorption chiller works using a proven design principle. It consists of four chambers: the evaporator, the condenser and two absorption chambers.

An adsorption chiller is a highly reliable, efficient solution that takes heat to produce chilled water, which in turn can be used for cooling and linked into refrigeration applications. Absorption technology produces chilled water down to as low as 3°C, as required, making it ideal for any application with a high cooling demand.

The two absorption chambers contain heat exchangers packed with silica, with a valve between them to allow vapour pressure equalization. Between each absorption chamber and the 'evaporator' and 'condenser' chambers, valves allow the transfer of vapour when there is sufficient pressure differential.

5.1.4 Distribution network

The distribution network operates in a four-pipe configuration (independent circuits for hot and cold water).

Figure 6: Absorption chillers at Geolit.

The 'pre-insulated' pipeline was selected for this installation due to the advantages of such pipes in front of *in situ* insulated pipes: minimal heat loss; quick installation; long life and minimum maintenance; and minor civil works (pipe buried directly).

This is a factory-made pre-insulated pipe, designed specifically for transporting hot or cold fluids in long distance systems, minimizing heat losses. These pipes are fitted with a rigid polyethylene outer shell that provides the mechanical protection needed for it to be directly buried in the ground.

5.2 Fuel and heat demand

The annual biomass consumption is approximately 1,800 tonnes, or 1,500 tonnes dry basis. In total, 7,674 MWh is the yearly energy demand for Geolit. The installation operates approximately 4,000 h per year in total. The boiler efficiency is 85%. Calculations show that the nominal heating capacity of Geolit is 3.26 MW.

5.3 Main advantages

The main advantages of this DHC system are as follows:

- It provides users with significant financial savings.
- The energy is directly delivered to the residents, as energy-producing elements in each building are not necessary. Consequently, it avoids risks and dirt, improves the use of spaces and eliminates maintenance or repair operations.
- It improves the energy efficiency and causes less environmental impact because the CO_2 emissions are avoided, unlike when generated with other systems: 0.63 kg CO_2/kWe for cold water production systems based on electricity; 0.21 kg CO_2/kWe for hot water production from natural gas.

6 Conclusions

The development of biomass as an energy resource makes its use for thermal energy production necessary. Among thermal applications, DHC systems are destined to play a crucial role because of their advantages for the environment but also for promoters, consumers and public administrations.

Nevertheless, the proper implementation of these systems requires:

- An improvement of the quality of biofuels.
- Development of the supply and logistic infrastructures.
- Higher stability of the biofuel price.
- Social concern.
- Technical developments focussed on the improvement of the boilers, which should increase their efficiency and make them capable of working with different types of biomass to ensure the energy supply.

References

[1] Instituto Catalán de la Energía y Asociación de Empresas de Redes de Calor y Frio, Guía básica de redes de distrito de calor y frío, Barcelona: Generalitat de Catalunya. 2011.

[2] European Union, Directive 2010/31/CE on the energy performance of buildings.

[3] European Union, Energy efficiency directive 2012/27/EU.

[4] District Heating & Cooling Platform, District heating & cooling strategic research agenda, 2012.

[5] La Cal-Herrera, J.A., Jurado, F. & Ogayar, B., A new model of energetic valuation for olive grove by-products based on gasification technologies integrated in an olive-mill. *International Journal of Green Energy*, **9(7)**, pp. 661–672, 2012.

[6] García-Munté Energía, S.L., www.garciamunte.com

[7] Eneragen, *Manual de promoción de ESEs*. IDAE: Madrid, 2010.

[8] European Union, Directive 2006/32/CE on energy end-use efficiency and energy services.

[9] BioSolESCo, ESCO – general issues, 2014.

[10] ECCJ, ESCO introduction and promotion study, 2006.

[11] EUBIONET III, Solutions for biomass fuel market barriers and raw material availability; WP6 – heating and cooling with biomass, 2010.

[12] IDAE, Biomasa: Climatización, IDAE: Madrid, 2009.

CHAPTER 10

Benefits from the Use of Pellet Boilers in the Energy Rating of Buildings

M. Carpio[1] and A. Garcia-Maraver[2]
[1]*Department of Building Construction, University of Granada, Spain.*
[2]*Department of Civil Engineering, University of Granada, Spain.*

Abstract

Using renewable energy contributes to reducing CO_2 emissions and improves energy ratings. The building sector is responsible for a significant portion of global CO_2 emissions. European Directive 2002/91/EC on the Energy Performance of Buildings, and the recast in Directive 2010/31/EU, configures the legislative framework of the European Union with respect to the energy rating of buildings in Member States.

This chapter summarizes the different transpositions regarding energy ratings and renewable energy sources, and more specifically it examines the CO_2 emissions and the energy rating using pellet boilers with different climatic conditions, based on real cases. It is concluded that replacing gasoil by wood pellets can lead to reductions in CO_2 emissions that range from 82.91% to 95.28%. Such a substantial reduction of CO_2 emissions directly affects the final energy rating, improving it as much as four levels on a scale of seven levels. Meanwhile, the economic savings from using wood pellets instead of gasoil may reach 68.82%.

Keywords: Energy certification, energy rating, wood pellets, buildings, CO_2 emissions.

1 Introduction

Global warming is directly linked with the emission of greenhouse gases, which include CO_2, CH_4 and N_2O. In recent years, their emissions have grown exponentially, leading to important environmental problems [1], and underlining the need to control their impact [2].

Buildings dedicated to living quarters are responsible for 40% of the energy consumed and 36% of the CO_2 emissions to the atmosphere in Europe [3,4].

Therefore, normative regulations – Directive 2002/91/CE [3], recast in the Directive 2010/31/EU [4] – were necessary to reduce the environmental impact generated by the building sector. The energy rating of buildings and their systems of heating and cooling, from the final users' standpoint, set energy requirements for a comfortable interior temperature, reduced emissions of CO_2 to the atmosphere and economic savings. In addition, governmental energy strategies include provisions for future grants for those residences that have optimal energy ratings.

The European Union, through its Directive 2002/91/CE [3], later recast in Directive 2010/31/EU [4], introduced energy performance certification for newly constructed buildings. This is one of the initiatives of the European Union against climate change deriving from the Kyoto Protocol [2], intended to reduce the environmental burden of emissions from the use of fossil fuels [1,2]. Directive 2002/91/CE replaces previous Directive 92/42/CEE regarding boilers, Directive 89/106/CEE regarding the products of construction and the provisions related with buildings corresponding to the program SAVE [5,6]. Each country is responsible for incorporating the guidelines specified in Directive 2002/91/CE and recast in Directive 2010/31/EU into the domestic legislative framework. The Directives formulate minimum requirements on energy performance and introduce a system of energy performance certification for buildings. They also call for developing plans for low or zero carbon buildings, with the public sector leading the way.

However, little attention was paid during their design or construction to the thermal performance of buildings, meaning that a very significant percentage would fail current energy examinations. For instance, over 50% of installed boilers run on fossil fuels [7]. Given the need to reduce the CO_2 emissions, the use of renewable fuels such as biomass should be encouraged. At present, 80% of the world energy is supplied by fossil fuels and 14% comes from renewable sources, with 9.6% thereof coming from traditional biomass [8]. This is an economically favourable alternative [9,10], which makes it possible to obtain beneficial energy ratings for existing buildings.

This chapter focuses on the impact of using pellet boilers on the energy rating and CO_2 emissions of residential buildings. Related studies using thermal simulations have been conducted in a number of countries under diverse conditions. For example, Pisello *et al.* [11] evaluated the influence of the climatic zone on the energy rating of buildings. Buratti *et al.* [12] concluded that building orientation and glazing systems improve thermal comfort and reduce the energy demand up to 67% in non-residential buildings. Studies in China [13], Spain [14] and the United Kingdom [14] have examined the energy efficiency performance of buildings using renewable energy sources for heating and domestic hot water (DHW), including biomass [13,14] and solar DHW [15]. Wang *et al.* [15] applied passive design methods and advanced façade designs to minimize the load requirements for heating and cooling purposes through building energy simulations, and analysed the local climate data. All these studies looked at factors affecting energy efficiency separately. The chapter at hand compares different parameters, including different climatic zones, and conventional and renewable fuels, such as wood pellets, for heating and DHW in different types of dwellings, with regard to: (1) energy demand; (2) environmental effects; (3) economic effects and (4) energy rating.

Given this background, and based on data reported previously [16], our main objective at this time is to determine the environmental and economic advantages of using wood pellets in systems for heating and DHW (as opposed to conventional energy sources), with reference to the energy rating of residential buildings. The methodology used, furthermore, makes it possible to determine the variables that bear the greatest influence on the energy rating of a building, and how the use of wood pellet biomass can contribute to an improved rating.

2 Legal Framework

The energy rating of a building is determined by theoretical thermal simulations that depend on the characteristics of the building and the classification of the climatic zone where it is located.

In general, the method of calculation is very similar in all European countries, taking the annual energy demand of the building to calculate the energy rating [17]. In the case of Sweden, calculation is based on the real quantity of energy used [18–20]; and other countries use a combination of the two methods for building energy ratings [17].

Calculating the energy rating according to the annual energy demand of the building requires great precision in the definition of the building envelope, materials, thermal bridges, heating and cooling, DHW, etc. This method is based on prediction [17]. Thus, it has the advantage of determining how the building will work before its actual use under normal conditions. In turn, calculating the real amount of energy used implies measurements that may vary among identical buildings in the same climate zone, because the human factor is involved in the calculation method [21], although a more individualized result for each dwelling is obtained.

According to the transposition of Directive 2002/91/CE [3], and the subsequent Directive 2010/31/EU [4], the calculation is to be carried out in the project phase. The objective of these Directives is to promote improvement of the energy performance of buildings (EPBD), taking into account outdoor climatic and local conditions, as well as indoor climate requirements and cost-effectiveness.

Not all European Union countries have adopted the same scale. There are scales with seven levels (used by Bulgaria [22], Cyprus [23], Czech Republic [24], France [25], Malta [26–28], Romania [29], Sweden [18–20], United Kingdom [30–34] and Norway [35]), to scales of up to 17 levels (e.g. Belgium – Brussels Capital Region [36]). Still other countries or regions have adopted a continuous scale (Belgium – Flemish Region [37], Germany [38,39], Latvia [40] and Portugal [41–43]).

The scale used in this chapter is the one most commonly used in Europe and comprises seven levels. The most efficient level is denoted by A, and the least efficient one designated by G (Fig. 1).

Figure 1: Energy rating label. Scale of seven levels.

Table 1 shows the EPBD transpositions to the different European countries. Some countries have a single regulation, whereas others have several additional transpositions.

3 Calculation Method in the Energy Rating

3.1 Climatic zoning

The climatic zone is defined as an area for which common external conditions for calculating the energy demand are defined using a few parameters [70]. The assignment of a correct climate zone is crucial, because the building faces different requirements depending on the climate zone, which affect the final energy rating. The classification of climatic zones chosen for use in this chapter to study the benefits of pellet boilers in the energy rating is a variation of the Köppen classification [71]. It involves the assignment of five different climatic zones for winter and four different climatic zones for summer [70]. The winter climates are designated by a letter (A, B, C, D and E) corresponding to the winter climate severity (WCS), whereas a number (1, 2, 3 and 4) represents the summer climate severity (SCS). As

Table 1: EPBD transpositions [17].

Country	EPBD transposition
Austria (AT)	Energy Performance Certificate Law (EAVG) [44]
Belgium – Brussels Capital Region (BE BR)	Brussels Air, Climate and Energy Code (BE) [36]
Belgium – Flemish Region (BE FR)	Execution Order of May 11, 2005, adopted in 2009 [37]
Belgium – Walloon Region (BE WR)	Calculation Procedures and Minimum Requirements for New and Existing Buildings [45], Certification of New Buildings [46], Certification of Existing Residential Buildings [47] and Certification of Existing Non-Residential Buildings [48]
Bulgaria (BG)	Energy Efficiency Act 2013 [22]
Croatia (HR)	Physical Planning and Building Act [49] and Energy Efficiency Act [50]
Cyprus (CY)	Law for the Regulation of the Energy Performance of Buildings [23]
Czech Republic (CZ)	Regulation on Energy Performance of Buildings [24]
Denmark (DK)	Danish Building Regulations (BR10) [51]
Estonia (EE)	Minimum Energy Performance Requirements [52]
Finland (FI)	National Building Code [53]
France (FR)	Energy Performance Diagnosis (DPE) [25]
Germany (DE)	Energy Saving Ordinance (EnEV) [39] and Renewable Heating Law (EEWärmeG) [38]
Greece (EL)	Law 3361 [54], KENAK (Regulation for Energy Performance of Buildings) [55], Presidential Decree 100/NG177 [56]
Hungary (HU)	Ministerial Decree on the Establishment of Energy Characteristics of Buildings [57] and Decree of Minister about Determination of Energy Efficiency of Buildings [58]
Ireland (IE)	Dwelling Energy Assessment Procedure (DEAP) and Non- Dwelling Energy Assessment Procedure (NEAP) [59]
Italy (IT)	Decree on the Promotion of the Use of Energy from Renewable Sources [60]
Latvia (LV)	Law on the Energy Performance of Buildings (LEPB) [40]
Lithuania (LT)	Law Energy Performance of Buildings [61]
Luxembourg (LU)	Grand-Ducal Regulation on the energy performance of buildings. Memorial and Functional [62]

(*Continued*)

Table 1: EPBD transpositions [17].

Country	EPBD transposition
Malta (MT)	Legal Notice of Minimum Requirements on the Energy Performance of Buildings [26], Legal Notice of Energy Performance of Buildings Regulations [27] and Legal of Energy Performance of Buildings Regulations [28]
Netherlands (NL)	Decree on Energy Performance of Buildings (BEG) [63] and Regulation on Energy Performance of Buildings (REG) [64]
Poland (PL)	Construction Act Journal [65]
Portugal (PT)	System of Energy Certification (SCE) [43], Regulation of Energy Systems and Climatization of Buildings (RSECE) [42] and Regulation of the Characteristics of Thermal Conduct of Buildings (RCCTE) [41]
Romania (RO)	Law of Energy Performance of Buildings [29].
Slovak Republic (SK)	Act on the Energy Performance of Buildings and on Amendment and Supplements to Certain Acts [66]
Slovenia (SI)	Regulation on Energy Performance [67]
Spain (ES)	Basic Procedure for Certification of Energy Efficiency of Buildings [68], Regulation of Thermal Installations in Buildings (RITE) [69] and Technical Code of Edification (CTE) [70]
Sweden (SE)	Law on Energy Declaration of Buildings [19], Performance Certificates for Buildings Ordinance [18] and Regulations by the National Board of Housing, Building and Planning [20]
United Kingdom – England and Wales (UK – EW)	Building Regulations (amendments) Regulations [30] Energy Performance of Buildings [31]
United Kingdom – Northern Ireland (UK – NI)	Building Regulations [32] and Energy Performance of Buildings (Certificates and Inspections) [33]
United Kingdom – Scotland (UK – S)	Building Act 2003, Building Regulations 2004, Building Procedure and Forms 2007, Energy Performance of Buildings Regulations 2008 [34]
Norway (NO)	Criteria for Passive Houses and Low Energy Buildings [35]

Table 2: Climatic zones [70].

		Summer Climate Severity (SCS)			
		1 SCS ≤ 0.6	2 0.6 < SCS ≤ 0.9	3 0.9 < SCS ≤ 1.25	4 SCS > 1.25
Winter Climate Severity (WCS)	A WCS ≤ 0.3	A1	A2	A3	A4
	B 0.3 < WCS ≤ 0.6	B1	B2	B3	B4
	C 0.6 < WCS ≤ 0.95	C1	C2	C3	C4
	D 0.95 < WCS ≤ 1.3	D1	D2	D3	D4
	E WCS > 1.3	E1	E2	E3	E4

shown in Table 2, the different combinations of these intervals amount to a total of 20 possible climatic zones. Yet, some are hardly feasible and could not be identified in Europe – for instance, an Antarctic climate or a Sahara desert climate [70]. Table 2 shows the thresholds of WCS and SCS.

3.2 Energy efficiency indicators method

The estimation of the energy necessary to comply with the demands of a building under normal conditions of occupancy and functioning is known as the Energy Efficiency Rating. By comparing a number of indicators of the mean energy use in model buildings of reference, a real building can be qualified and certified on an energy scale established for this purpose [68,72].

The Energy Efficiency Indicators (EEI) in residential buildings are: (1) EEI heating demand; (2) EEI cooling demand; (3) EEI of heating emissions; (4) EEI of cooling emissions; (5) EEI of emissions for DHW; and (6) EEI of total emissions.

3.2.1 Energy efficiency indicator heating demand

It is the ratio between the heating demand of the studied building and the reference heating demand. For residential buildings, the heating demand is the reference corresponding to the average value of similar new buildings that conform with the regulations of a given year (in this case 2006).

This mean value depends on the locality in which the building is located. It is different for single-family houses and residential buildings.

3.2.2 Energy efficiency indicator cooling demand

This is the ratio between the cooling demand of the studied building and the reference cooling demand. In the case of residential buildings, the cooling demand is the reference corresponding to the average value of similar new buildings in conformity with the regulations in a given year (in this case 2006).

This mean value depends on the locality in which the building is located, and it is also different for single-family houses and residential buildings.

3.2.3 Energy efficiency indicator of heating emissions

It is the ratio of CO_2 emissions due to heating service in the studied building and CO_2 emissions of reference for the heating service.

3.2.4 Energy efficiency indicator of cooling emission

It is the ratio of CO_2 emissions due to cooling service in the studied building and CO_2 emissions of reference for the cooling service.

3.2.5 Energy efficiency indicator emission for DHW

This is the ratio of CO_2 emissions due to DHW service in the studied building with respect to CO_2 emissions of reference for the DHW service.

3.2.6 Energy efficiency indicator of total emissions

It is the ratio between the total CO_2 emissions caused by all the services considered in the building object and total CO_2 emissions of reference for the same services. Total CO_2 emissions of the building as well as the building object reference are obtained by adding the CO_2 emissions for each service considered.

3.3 Method for obtaining efficiency classes

The rate of energy rating, C1, is obtained from the value of the indicator of energy efficiency (IEE) by

$$C_1 = \frac{(\mathrm{IEE} \times R) - 1}{2 \times (R - 1)} + 0.6$$

where R is the ratio between the value of the indicator for the percentile 50% and the percentile 10% of new residential buildings of 2006 according to the housing census.

Table 3 shows the values of R (dispersion of the IEE, to use in total emissions).

The limits of the scale are expressed through the energy rating index C1, based on Table 4.

This scale comprises seven levels, the most efficient one denoted by A, and the least efficient one designated by G. No new buildings would have levels F or G, as these are used only for renovated structures [74].

Table 3: Values of *R* [73].

		Summer climate zone			
		1	2	3	4
Winter climate zone	A			1.60	1.60
	B			1.60	1.55
	C	1.50	1.50	1.55	1.55
	D	1.45	1.50	1.50	
	E	1.45			

Table 4: Limits of efficiency classes [73].

Level	Limits	
A	C1	<0.15
B	0.15 ≤ C1	<0.50
C	0.50 ≤ C1	<1.00
D	1.00 ≤ C1	<1.75
E	C1	>1.75

3.4 Simulation software

The use of software designed for calculating the energy rating began in the 1980s [75]. It originally involved theoretical thermal simulations, and eventually sophisticated tools arose, including exhaustive climate records, libraries of materials with different constructive solutions and complete CAD integration.

The various programs available nowadays differ in terms of how the characteristics of the building are introduced as input, and in the output provided [76]. The software chosen for this study was CERMA [77], which works on the scale of seven levels.

4 Case Study

4.1 Characteristics

4.1.1 Fuel characteristics

Biomass is defined as 'all biological materials excluding those that were included in geological formations suffering a mineralization process' [78]. Accordingly, coal, oil and natural gas are excluded. For the purposes of this study, wood pellets were selected as the fuel [79], and for comparison with fossil fuels, we chose natural gas and gasoil. Table 5 shows the lower heating value (LHV), density and price of fuels used for the study.

Table 5: Fuel characteristics.

Fuel	LHV	Density	Price
Gasoil [80]	11.89 kWh/kg	850 kg/m^3	1.100 €/l–0.109 €/kWh
Natural gas [81]	11.63 kWh/m^3	*n/n*	0.059 €/kWh
Wood pellet [82]	5.01 kWh/kg	*n/n*	0.034 €/kWh

n/n, not necessary for this study.

The present chapter focuses on heating and DHW since this system uses fuel directly (gasoil, natural gas or wood pellet). Because the energy rating procedure calls for choosing a system of refrigeration as well, also considered is an electrical based refrigeration system, which would be the most commonly used system in residential buildings.

4.1.2 Construction characteristics

Two types of buildings were selected: (1) a single-family house and (2) a multi-family residential building, placed among other constructions.

The single-family dwelling consists of three floors with total usable area 261.99 m^2: a basement (108.73 m^2), a ground floor (117.01 m^2) and a first floor (36.25 m^2). The house is located on a gentle slope, which means that the basement is completely underground on one side, yet above the ground on the other side of the house.

The multi-family dwelling has five stories with total usable area 861.71 m^2: a ground floor (267.10 m^2), a first (267.10 m^2), a second (267.10 m^2), a third floor (267.10 m^2) and a tower (18.96 m^2). In this case, the building is a rectangle on a corner so that the north and east sides of it are fully in contact with other constructions, while the south and west façades are exposed.

The most important materials in the thermal enclosure and thermal transmittance limit (U) used were: roof (0.48 W/m^{2}k), uninhabitable area roof (0.75 W/m^{2}k), external wall (0.54 W/m^{2}k), ground floor (0.65 W/m^{2}k), wood door (2.20 W/m^{2}k), garage door (3.20 W/m^{2}k) and windows (2.47 W/m^{2}k). The main façade was facing south in all cases. Continuous insulation implies junctions with framework slab, and constant closure to the line of the doorjamb, lintel or windowsill, due to the thermal bridges, given their role in heat loss. For instance, inadequate execution of exterior closures of a double brick wall can mean 30% more thermal losses [83].

For the thermal simulation of each building and climatic zone, boilers with similar characteristics – able to fire gasoil, natural gas or wood pellet – were chosen. The thermal load selected for each boiler was set to 24 kW. For the single-family house, just one boiler (24 kW) was considered, whereas for the multi-family dwelling three boilers were installed (total boiler load of 72 kW). For all boilers, the thermal efficiency value adopted was 90%, with an outlet water temperature of

50°C for DHW and 80°C for heating. The flow rate of DHW in the single-family house was 235.80 L/day, and in the multi-family dwelling 568.72 L/day. Both types of residence featured an accumulator; specifically, it had a capacity of 200 L in the single-family house and 500 L in the multi-family dwelling. In both cases, the water temperature varied between 60°C and 80°C, and the global heat transfer coefficient (U×A) being 1 W/K.

4.1.3 Climatic zones

For this study, the most common climatic zones were selected, including extremes zones (the warmest, A4 and B3; the coldest, D2 and E1) and intermediate zones (C4 and C3). Table 2 shows the different climatic zones.

4.1.4 Indoor temperature

The indoor temperature of the residences is determined by the climate, season and the heating/cooling system used. The colder climate zones show fairly even temperatures in all months of the year, except during summer, revealing that heating systems provide a very stable indoor temperature in winter (between 17°C and 20°C). During summer, temperatures are somewhat irregular since there is no need for cooling, with a mean temperature of 21°C and a maximum of 24°C. The lowest temperatures, in May and June, can be attributed to an interruption in the use of heating together with outdoor temperatures generally lower than 17°C. In contrast, the dwellings situated in the warm climate zones show very irregular temperature during winter: the outdoor temperature often reaches 22°C to 23°C, meaning that heating is not required, whereas during summer the indoor temperatures are regulated by the usual use of a cooling system.

4.2 Results

4.2.1 Energy demand

Table 6 shows the energy demand in the buildings and different climatic zones. The energy demand data obtained through simulations of ideal and equivalent situations indicate the objectives to attain in the blueprint stage. Once a residence is occupied, the 'user factor' can significantly affect the results, depending on the residents' particular habits, and their maintenance and use of the home. For example, two adjacent and identical dwellings may show up to 40% variability in their heating expenses due to excessive ventilation [21]. This implies that real data may vary from 50% to 150% with regard to the theoretical calculations [84].

All the houses studied share the same essential features, so that the only factor influencing the energy demand is the climatic zone, which has a great impact on the results. Table 6 reveals that the total energy demand ranges from 55.7 kWh/m^2 year in A4 to 164.1 kWh/m^2 year in B1 for single-family houses, and from 44.7 kWh/m^2 year in A4 to 136.5 kWh/m^2 year in B1 for multi-family residences. The variations are particularly high in the case of cities with harsher climates, where the heating demand is greater [10,15].

Table 6: Energy demand (kWh/m^2 year).

	Single-family house					
Climatic zone	A4	B3	C4	C3	D2	E1
Heating	13.7	32.5	47.2	62.3	76.5	143.7
Cooling	23.8	12.9	18.3	11.5	9.9	
DHW	18.2	18.7	18.8	18.8	19.2	20.4
Total	55.7	64.1	84.3	92.6	105.6	164.1
	Residential building					
Climatic zone	A4	B3	C4	C3	D2	E1
Heating	17.1	35.5	46.8	60.9	72.0	121.2
Cooling	14.0	7.9	9.3	6.8	6.2	
DHW	13.6	14.0	14.1	14.0	14.4	15.3
Total	44.7	57.4	70.2	81.7	92.6	136.5

It is also observed that the single-family house uses 20.22%–24.61% more energy than the multi-family house, depending on the climatic zone. In fact, the enclosure of a building (m^2) and its volume (m^3) are 26% greater for the single-family residence, which means larger exposure to the elements.

There is a progressive increase in energy demand from warmer to colder areas. The heating demand in B1 is 1048.9% greater than that in A4 for a single-family house, and 708.8% greater for a multi-family house. It may be concluded that energy demand for heating is inversely proportional to the winter outdoor temperatures.

In the case of cooling, A4 is the climatic zone with the greatest demand, requiring 204.4% more energy than the single-family residence in D2, and 225.8% more than the multi-family house. B1 was not included in this aspect of the study since it does not need cooling in the summer, when the outdoor temperature remains within the comfort zone. Hence, there is a progressive increase in the energy demand for cooling related to higher temperatures.

Finally, regarding DHW, the demand appears to depend largely upon the area of the living quarters. The influence of the climatic zone is minimal, giving differences between the two cities with extreme climates of 12.1% for the single-family unit and 12.5% for the multi-family unit.

4.2.2 CO_2 emissions

Table 7 also shows the CO_2 emissions, expressed in kg CO_2/m^2 year, released to the atmosphere by the residential units as a consequence of the energy demands. The emissions due to the heating systems are much higher for climatic zone E1, regardless of the type of house. The values range from 0.5 kg CO_2/m^2 year using wood pellets in the single-family house in A4 to 51.1 kg CO_2/m^2 year for the single-family house in E1 using gasoil. The single-family house shows an increase

Table 7: CO_2 emissions (kg CO_2/m^2 year).

		Single-family house						Residential building					
Fuel	Climatic zone	A4	B3	C4	C3	D2	E1	A4	B3	C4	C3	D2	E1
Gasoil	Heating	4.6	10.7	16.1	22.0	26.3	51.1	5.7	11.8	15.6	20.3	24.2	41.8
	Cooling	9.1	4.9	7.0	4.4	3.8	–	5.3	3.0	3.5	2.6	2.4	–
	DHW	6.0	6.1	6.2	6.1	6.3	6.6	4.5	4.6	4.6	4.6	4.7	5.0
	Total	19.7	21.7	29.3	32.5	36.4	57.7	15.5	19.4	23.7	27.5	31.3	46.8
Natural gas	Heating	3.4	7.9	12.1	16.6	19.8	38.9	4.3	8.9	11.8	15.3	18.4	32.1
	Cooling	9.1	4.9	7.0	4.4	3.8	–	5.3	3.0	3.5	2.6	2.4	–
	DHW	4.2	4.4	4.4	4.4	4.5	4.7	3.2	3.3	3.3	3.3	3.3	3.5
	Total	16.7	17.2	23.5	25.4	28.1	43.6	12.8	15.2	18.6	21.2	24.1	35.6
Wood pellet	Heating	0.5	1.0	2.1	3.4	3.6	8.8	0.9	1.9	2.4	3.2	4.1	8.0
	Cooling	9.1	4.9	7.0	4.4	3.8	–	5.3	3.0	3.5	2.0	2.4	–
	DHW	0	0	0	0	0	0	0	0	0	0	0	0
	Total	9.6	5.9	9.1	7.8	7.4	8.8	6.2	4.9	5.9	5.2	6.5	8.0

of CO_2 emissions of 1760.0% with wood pellets as opposed to gasoil; in the multi-family residential building, the increase comes to 888.8%.

Gasoil is the fuel that releases more CO_2 during its combustion for the purpose of heating and DHW [10], while wood pellets prove to be the most favourable fuel from the standpoint of CO_2 emissions [10,85,86].

Table 8 compares the CO_2 emissions from systems using gasoil and wood pellets. It is seen that replacing gasoil by wood pellets leads to reductions in the CO_2 emissions that range from 82.91% to 95.28%. Similar results were obtained by Pardo and Thiel [10], who report reductions of around 95% in CO_2 emissions using biomass for Southern Europe when compared with conventional fossil fuel fired systems.

The present chapter reveals that the replacement of gasoil by any other fuel for heating and DHW purposes reduces the CO_2 emissions, although the use of wood pellets is the most favourable.

It is noteworthy that the differences in CO_2 emissions from one climatic zone to another depend not only on the energy demand but also on the fuel type and the dwelling area.

An increase in household size tends to decrease the energy consumption per unit area, i.e. the CO_2 emission density or CO_2 emission per unit area, which is commonly used worldwide to determine the building energy efficiency rating [87]. In summary, the smaller the dwelling area is, the lower the building energy efficiency rating can be set. Meanwhile, the larger the dwelling area is, the higher the energy efficiency rating can be set [87].

4.2.3 Energy rating

Based on existing legislation, the rating interval determined for the buildings under consideration is based on a calculation with respect to a model building and other

Table 8: CO_2 emission savings: pellet versus gasoil.

	Single-family house					
Climatic zone	A4	B3	C4	C3	D2	E1
Heating	89.13%	90.65%	86.96%	84.55%	86.31%	82.78%
H + DHW	95.28%	94.05%	90.58%	87.90%	88.96%	84.75%
H + DHW + cooling	51.27%	72.81%	68.94%	76.00%	79.67%	84.75%
	Residential building					
Climatic zone	A4	B3	C4	C3	D2	E1
Heating	84.21%	83.90%	84.62%	84.24%	83.06%	80.86%
H + DHW	91.18%	88.41%	88.12%	87.15%	85.81%	82.91%
H + DHW + cooling	60.00%	74.74%	75.11%	81.09%	79.23%	82.91%

H, Heating; DHW, domestic hot water.

Table 9: Energy rating in the buildings and climatic zones studied.

	Single-family house					
Climatic zone	A4	B3	C4	C3	D2	E1
Gasoil	D	D	D	D	D	D
Natural gas	D	D	D	D	D	D
Wood pellet	C	B	B	A	A	A
	Residential building					
Climatic zone	A4	B3	C4	C3	D2	E1
Gasoil	E	E	E	E	E	E
Natural gas	D	D	D	D	D	D
Wood pellet	C	B	B	B	A	A

references. Table 9 shows the energy rating for the buildings and climatic zones studied. The threshold limits of the levels of energy ratings are seen to depend on the type of residence, climatic zone and fuels used.

Improvement in the energy rating of a building is directly related to the fuel type. Gasoil and natural gas imply the assignment of rating D for single-family dwellings, and E for multi-family units in the six climatic zones studied here. In the case of pellets, as well as for all biomass, the rating depends on climate, but is independent of the housing type, with improvements associated with the lower winter mean temperatures, which may result in upgrading of up to four levels, that is, C would be the rating in the case of the hottest city, A in the coldest three cities and B for the remaining cases.

4.2.4 Economy

To determine the costs involved in using the heating systems with the different fuels, an economic analysis was carried out. Bearing in mind the fuel costs (Table 5) and the energy demand (Table 6), costs were evaluated for heating and DHW together. Table 10 summarizes the results of this analysis. Costs were found to be directly related to the energy demand.

Regarding the most economical climatic zone, A4, the following results were obtained, regardless of the fuel used: (1) for single-family unit, costs in C4 were 60.50% higher, in B3 106.90%, in C3 154.30%, in D2 300.00% and in E1 414.42% higher and (2) for the multi-family unit, costs in B3 were 61.24% higher, C4 98.37%, C3 143.97%, D2 181.43% and E1 344.62% higher. It should be stressed that savings of 68.82% are achieved by changing from gasoil to wood pellets. The use of natural gas instead of gasoil yields savings of 54.21%. Therefore, it may be concluded that the most economic fuel is generally the wood pellet. Other studies arrived at savings of ≈95% in Central and Northern Europe and ≈75% in the case of Southern European regions in comparison with conventional systems [10].

Table 10: Economic cost: heating + DHW.

	Single-family house					
Climatic zone	A4	B3	C4	C3	D2	E1
Gasoil	838 €	1,345 €	1,734 €	2,130 €	2,514 €	4,310 €
Natural gas	454 €	729 €	940 €	1,155 €	1,363 €	2,336 €
Wood pellet	261 €	419 €	540 €	664 €	783 €	1,344 €
	Residential building					
Climatic zone	A4	B3	C4	C3	D2	E1
Gasoil	2,487 €	4,011 €	4,934 €	6,069 €	7,000 €	11,060 €
Natural gas	1,348 €	2,174 €	2,675 €	3,290 €	3,795 €	5,995 €
Wood pellet	775 €	1250.36 €	1538.32 €	1891.96 €	2182.44 €	3,448 €

5 Conclusions

The transposition of the European framework to each country has given rise to a series of regulations with a common origin that are nonetheless not uniform or homogeneous among them.

In view of the current transpositions, it is impossible to compare the energy efficiency of two identical buildings in different countries, even when the same climatic conditions prevail, because the energy scales are different, as are the calculation methods (energy demand, real consumption or both).

Based on the most commonly applied method in Europe, for an energy rating with a scale of seven levels, this chapter may conclude that the use of wood pellets in heating and DHW residential systems presents three important advantages: (1) the environmental costs are reduced, since significantly less CO_2 is released; (2) a very favourable energy rating is ensured and (3) important economic savings are provided.

The CO_2 emissions depend directly on the climatic zone where the house is located, in addition to the fuel used. Gasoil was found to yield higher CO_2 emissions regardless of the housing type. However, the use of pellets instead of gasoil or natural gas was found to bring about an important reduction of the CO_2 emissions in all cases. Specifically, if gasoil is replaced by pellets, reductions in CO_2 emissions as high as 95.25% for single-family units, and 91.18% for multi-family units, are achieved. Bearing in mind that 40% of the energy consumed in Europe and 36% of the CO_2 emissions to the atmosphere are produced by buildings dedicated to living quarters, the choice of fuel stands as a significant factor in evaluating emissions derived from heating and DHW.

Using wood pellets for heating purposes enhances the energy rating of the housing units in all cases. In the best case scenario, improvements are of four points on the scale of residential energy performance, which would put the unit into the top category, A. In comparison, with the use of fossil fuels, the best rating is D for the single-family residence and E for the multi-family one.

Cost-effectiveness is another important area where savings by means of wood pellets are noteworthy. In comparison with gasoil, the use of pellets can lead to economic savings of as much as 70%.

In short, the use of pellets is advantageous in all respects, providing for environmental and economic benefits, while leading to a better energy rating.

References

[1] Petit, J.R., Jouzel, J. & Raynaud, D., Climate and atmospheric history of the past 420,000 years from the Vostok ice core, Antarctica. *Nature*, **399**, pp. 429–436, 1999.

[2] United Nations, Kyoto protocol to the United Nations framework convention on climate change, 1997.

[3] European Parliament and of the Council, Directive 2002/91/EC of the European Parliament and of the Council of 16 December on the energy performance of buildings. *DOUE*, **1**, pp. 65–71, 2003.

[4] European Parliament and of the Council, Directive 2010/31/EU of the European Parliament and of the Council of 19 May on the energy performance of buildings. *DOUE*, **153**, pp. 13–35, 2010.

[5] European Parliament and of the Council, Multiannual programme for the promotion of energy efficiency (SAVE) (1998 to 2002). Decision 647/2000/-EC of 28 February. *DOUE*, **79**, pp. 6–9, 2000.

[6] European Parliament and of the Council, Multiannual programme for the promotion of energy efficiency in the Community (SAVE II). Decision 96/737/EC of 16 December, Annual report, 1996.

[7] Eurostat European Commission, Ministerio de Industria Energía y Turismo, Government of Spain, Instituto para la Diversificación y Ahorro de la Energía. Proyecto SECH-SPAHOUSEC. Analisis del consumo energético del sector residencial en España. Final ed., Madrid, 2011.

[8] Khan, A.A., de Jong, W., Jansens, P.J. & Spliethoff, H., Biomass combustion in fluidized bed boilers: potential problems and remedies. *Fuel Processing Technology*, **90**, pp. 21–50, 2009.

[9] Abulfotuh, F., Energy efficiency and renewable technologies: the way to sustainable energy future. *Desalination*, **209**, pp. 275–282, 2007.

[10] Pardo, N. & Thiel, C., Evaluation of several measures to improve the energy efficiency and CO_2 emission in the European single-family houses. *Energy and Buildings*, **49**, pp. 619–630, 2012.

[11] Pisello, A.L., Goretti, M. & Cotana, F., A method for assessing buildings' energy efficiency by dynamic simulation and experimental activity. *Applied Energy*, **97**, pp. 419–429, 2012.

[12] Buratti, C., Moretti, E., Belloni, E. & Cotana, F., Unsteady simulation of energy performance and thermal comfort in non-residential buildings. *Building and Environment*, **59**, pp. 482–491, 2013.
[13] Cao, G., Jokisalo, J., Feng, G., Duanmu, L., Vuolle, M. & Kurnitski, J., Simulation of the heating performance of the Kang system in one Chinese detached house using biomass. *Energy and Buildings*, **43**, pp. 189–199, 2011.
[14] Ruiz, M.C. & Romero, E., Energy saving in the conventional design of a Spanish house using thermal simulation. *Energy and Buildings*, **43**, pp. 3226–3235, 2011.
[15] Wang, L., Gwilliam, J. & Jones, P., Case study of zero energy house design in UK. *Energy and Buildings*, **41**, pp. 1215–1222, 2009.
[16] Carpio, M., Zamorano, M. & Costa, M., Impact of using biomass boilers on the energy rating and CO_2 emissions of Iberian Peninsula residential buildings. *Energy and Buildings*, **66**, pp. 732–744, 2013.
[17] Carpio, M., García-Maraver, A., Ruiz, D.P., Martínez, A. & Zamorano, M., Energy rating for green buildings in Europe. *WIT Transactions on Ecology and the Environment*, **190(1)**, pp. 381–394, 2014.
[18] Ministry of Enterprise, Energy and Communications, Government of Sweden, Performance Certificates for Buildings Ordinance. *Ordinance* 2006:1592, 2006.
[19] Ministry of Enterprise, Energy and Communications, Government of Sweden, Law on Energy Declaration of Buildings. *Law* (2006:685), 2006.
[20] Ministry of Enterprise, Energy and Communications, Government of Sweden, National Board of Housing, Building and Planning, 2012.
[21] Zabalza, I., Díaz, S. & Aranda, A., *Manual práctico de certificación energética de edificios*. Universidad de Zaragoza: Zaragoza, 2009.
[22] Ministry of Economy and Energy, Government of Bulgaria, Energy Efficiency Act SG24/12 03 2013.
[23] Ministry of Energy, Commerce, Industry and Tourism, Government of Cyprus, Law for the Regulation of the Energy Performance of Buildings. L142(I)/2006, 2006.
[24] Ministry of Industry and Trade, Government of Czech Republic, Regulation on Energy Performance of Buildings. Regulation 148/2007, 2007.
[25] Ministry of Ecology and Sustainable Development Energy, Government of France, Diagnostic de Performance Énergétique (DPE). *décret no.* 2011-413 du 13 avril, JORF 0092, p. 6840, 2011.
[26] Ministry for Resources and Rural Affairs, Government of Malta, Minimum Requirements on the Energy Performance of Buildings. Legal Notice 238 of 2006.
[27] Ministry for Resources and Rural Affairs, Government of Malta, Energy Performance of Buildings Regulations. Legal Notice 261 of 2008.
[28] Ministry for Resources and Rural Affairs, Government of Malta, Energy Performance of Buildings Regulations. Legal Notice 376 of 2012.
[29] Ministry of Regional Development and Public Administration, Government of Romania, Law of energy performance of buildings. Law 372/2005, 2005.

[30] Welsh Government, Building Regulations (amendments) Regulations. Statutory Instrument 2012/3119, 2012.
[31] Welsh Government, Energy Performance of Buildings. Statutory Instrument 2012/3118, 2012.
[32] Department of Finance and Personnel Northern Ireland (DFPNI), Building Regulations. Statutory Rule 2012 No. 192, 2012.
[33] Department of Finance and Personnel Northern Ireland (DFPNI), Energy Performance of Buildings (Certificates and Inspections). Statutory Rule 2008 No. 170, 2008.
[34] Directorate for the Built Environment, Government of Scotland, Energy Performance of Buildings Regulations, 2008.
[35] Water Resources and Energy Directorate (NVE), Government of Norway, Criteria for passive houses and low energy buildings. NS 3701, 2012.
[36] Regional Ministry of Energy, Government of the Brussels Capital Region, Brussels Air, Climate and Energy Code. BE on 19 April 2013.
[37] Flemish Energy Agency (VEA), Government of Belgium, Execution Order of May 11, 2005, adopted in 2009.
[38] Federal Ministry for Environment, Nature Conservation and Nuclear Safety, Government of Germany, Erneuerbaren-Energien-Wärmegesetz (EEWärmeG), 2009.
[39] Federal Ministry of Transport, Building and Urban Development, Government of Germany, Energieeinsparverordnung für Gebäude (EnEV), 2009.
[40] Ministry of Economy, Government of Latvia, Ēku energoefektivitātes likums. Law on the Energy Performance of Buildings (LEPB), 2008.
[41] Ministério das Obras Públicas, Transportes e Comunicações, Government of Portugal, Regulamento das Características de Comportamento Térmico dos Edifícios (RCCTE). Decreto-Lei no. 80/2006 de 4 de Abril, DR 67, pp. 2468–2513, 2006.
[42] Ministério das Obras Públicas, Transportes e Comunicações, Government of Portugal, Regulamento dos Sistemas Energéticos e de Climatização nos Edifícios (RSECE). Decreto-Lei no. 79/2006 de 4 de Abril, DR 67, pp. 2416–2468, 2006.
[43] Ministério da Economia e da Inovação, Government of Portugal, Sistema de certificação energética (SCE). Decreto-Lei no. 78/2006 de 4 de Abril, DR 67, pp. 2411–2415, 2006.
[44] Austrian Institute of Construction Engineering (OIB), Energieausweis-Vorlage-Gesetz (Energy Performance Certificate Law). EAVG, 2012.
[45] Department of Energy and Sustainable Buildings, Government of Belgium, Calculation procedures and minimum requirements for new and existing buildings. MB du 22/06/2012, p. 34014, 2012.
[46] Department of Energy and Sustainable Buildings, Government of Belgium, Certification of new buildings. MB du 05/09/2011, p. 56370, 2011.
[47] Department of Energy and Sustainable Buildings, Government of Belgium, Certification of existing residential buildings. MB du 07/06/2010, p. 35958, 2010.

[48] Department of Energy and Sustainable Buildings, Government of Belgium, Certification of existing non-residential buildings. MB du 03/11/2011, p. 65830, 2011.
[49] Ministry of Construction and Physical Planning, Government of Croatia, Physical Planning and Building Act. Official Gazette No.76, 2007.
[50] Ministry of Construction and Physical Planning, Government of Croatia, Energy Efficiency Act. Official Gazette No. 152, 2008.
[51] Ministry of Business and Growth, Government of Denmark, Danish Building Regulations, BR10, 2010.
[52] Ministry of Economic Affairs and Communications, Government of Estonia, Minimum Requirements for Energy Efficiency. Decree nr. 258, 2009.
[53] Ministry of Environment, Government of Finland, National Building Code. NBD 2013, 2013.
[54] Ministry of Environment, Energy and Climate Change, Government of Greece, Transposition of the EPBD. Law 3661 (18 May 2008), 2008.
[55] Ministry of Environment, Energy and Climate Change, Government of Greece, KENAK (Regulation for Energy Performance of Buildings). Ministerial decision D6/B/5825, National Gazette 407, 2010.
[56] Ministry of Environment, Energy and Climate Change, Government of Greece, Presidential Decree 100/NG177. National Gazette 6th October, 2010.
[57] Ministry of Interior, Government of Hungary, Ministerial Decree on the establishment of energy characteristics of buildings. MD TNM 7/2006, 24, 2006.
[58] Ministry of Interior, Government of Hungary, Decree of Minister without Portfolio About Determination of Energy Efficiency of Buildings. Hungarian Decree 40/2012, 13, 2012.
[59] Department of the Environment, Community and Local Government (DECLG), Government of Ireland, Dwelling Energy Assessment Procedure (DEAP) and Non-Dwelling Energy Assessment Procedure (NEAP). SI 243 of 2012, 2012.
[60] Ministry for Economic Development, Government of Italy, Fourth Conto Energia. Decree 28/2011, 2011.
[61] Ministry of Energy, Government of Lithuania, Law on Energy, Energy Performance of Buildings. STR 2 01 09, 2005.
[62] Ministry of Economy and Foreign Trade, Government of Luxembourg, Règlement grand-ducal modifié du 31 août 2010 concernant la performance énergétique des bâtiments fonctionnels. A No. 173 de 2010, 2010.
[63] Ministry of the Interior and Kingdom Relations, Government of the Netherlands, Decree on Energy Performance of Buildings (BEG), 2006.
[64] Ministry of the Interior and Kingdom Relations, Government of the Netherlands, Energy Performance of Buildings (REG), 2009.
[65] Ministry of Infrastructure, Government of Poland, Construction Act Journal Laws No. 191, 1373, 2009.

[66] Ministry of Construction and Regional Development, Government of Slovakia, Energy Performance of Buildings and on Amendment and Supplements to Certain Acts. Act 555/2005, 2005.
[67] Ministry of the Economy, Energy and Mining Inspectorate, Government of Slovenia, Regulation on Energy Performance, 2010.
[68] Ministerio de la Presidencia, Government of Spain, Procedimiento básico para la certificación de la eficiencia energética de los edificios. Real Decreto 235/2013, de 5 de Abril, BOE 89, pp. 27548–27562, 2013.
[69] Ministerio de Industria, Turismo y Comercio, Government of Spain, Ministerio de la Vivienda, Government of Spain, Reglamento de Instalaciones Térmicas en los Edificios (RITE). Real Decreto 1027/2007 de 20 de julio, BOE 207, pp. 35931–35984, 2007.
[70] Ministerio de la Vivienda, Government of Spain, Código Técnico de la Edificación (CTE). Real Decreto 314/2006 de 17 de marzo, BOE 74, pp. 11816–11831, 2006.
[71] Chen, D. & Chen, H.W., Using the Köppen classification to quantify climate variation and change: an example for 1901–2010. *Environmental Development*, **6**, pp. 69–79, 2013.
[72] Ministerio de la Presidencia, Government of Spain, Procedimiento básico para la certificación de eficiencia energética de edificios de nueva construcción. Real Decreto 47/2007 de 19 de enero, BOE 27, pp. 4499–4507, 2007.
[73] Ministerio de la Vivienda, Government of Spain, Ministerio de Industria, Turismo y Comercio, Government of Spain, Instituto para la Diversificación y Ahorro de la Energía. Condiciones de aceptación de procedimientos alternativos a Lider y Calener, 2009.
[74] Ministerio de Industria, Turismo y Comercio, Government of Spain, Ministerio de la Vivienda, Government of Spain, Instituto para la Diversificación y Ahorro de la Energía. Escala de calificación energética. Edificios de nueva construcción. Madrid: IDAE, 2009.
[75] Newton, D., James, R. & Bartholomew, D., Building energy simulation – a user's perspective. *Energy and Buildings*, **10**, pp. 241–247, 1988.
[76] Crawley, D.B., Hand, J.W., Kummert, M. & Griffith, B.T., Contrasting the capabilities of building energy performance simulation programs. Building and Environment, **43**, pp. 661–673, 2008.
[77] Asociación Técnica Española de Climatización y Refrigeración, Universidad de Valencia, CERMA.V. 2.2, www.atecyr.org, 2011.
[78] AENOR, European Technical Specification. Solid biofuels. Terminology, definitions and descriptors. UNE-EN 14588, 2010.
[79] Van Loo, S. & Koppejan, J., *The Handbook of Biomass Combustion and Co-firing*, Earthscan: London, 2007.
[80] Ministerio de Industria Energía y Turismo, Government of Spain, Geoportal. http://geoportal.mityc.es.
[81] Comisión Nacional de la Energía, www.cne.es.
[82] Instituto para la Diversificación y Ahorro de la Energía, www.idae.es.

[83] Theodosiou TG, Papadopoulos AM, The impact of thermal bridges on the energy demand of buildings with double brick wall constructions. *Energy and Buildings*, **40**, pp. 2083–2089, 2008.

[84] AENOR, Thermal performance of buildings. Calculation of energy use for heating. Residential buildings. UNE-EN 832:2000, 2000.

[85] Dion, L., Lefsrud, M. & Orsat, V., Review of CO_2 recovery methods from the exhaust gas of biomass heating systems for safe enrichment in greenhouses. *Biomass and Bioenergy*, **35**, pp. 3422–3432, 2011.

[86] Joelsson, J. & Gustavsson, L., Swedish biomass strategies to reduce CO_2 emission and oil use in an EU context. *Energy*, **43**, pp. 448–468, 2012.

[87] Koo, C., Hong, T., Lee, M. & Seon Park, H., Development of a new energy efficiency rating system for existing residential buildings. *Energy Policy*, **68**, pp. 218–231, 2014.

Biomass to Biofuels

Edited by: ***S. SYNGELLAKIS****, Wessex Institute of Technology, UK*

Biomass is a continuously renewed source of energy formed from or by a wide variety of living organisms. Through biochemical and thermochemical processes, it is converted into gaseous, liquid or solid biofuels, which already meet a significant share of the current world energy needs. Because of their contribution to the sustainability of energy supply, reduction of greenhouse gas emissions as well as local employment and energy self-reliance, research interest and activity in enhancing biofuel energy output, efficiency and performance remain strong.

The first part of this volume comprises five articles mainly concerned with biomass resource potential and management. More specifically, the reported investigations assess grass and lawn substrates, rapeseed straw and microalgae from Upflow Anaerobic Sludge Blanket (UASB) reactor effluents as possible sources of biogas, bioethanol and biodiesel, respectively. The emphasis in the subsequent group of eleven articles is on biomass conversion processes, aiming at assessing performance as well as output quality and diversity. Biodiesel, a fluid biofuel produced from biomass with high lipids such as rapeseed oil, sunflowers and soy beans, is the focus of two articles: the first investigates the effect of biodiesel blending with diesel fuel on diesel engine performance and emissions; the second assesses the efficiency of catalytic reforming of biodiesel into a gaseous mixture used directly as Solid Oxide Fuel Cell (SOFC) fuel. The last three articles examine the prospects of biofuels as viable sources of energy within European contexts.

Biomass to Biofuels addresses a significant number of important themes and thus combines subject breadth and density with in-depth study of biomass resourcing and processing as well as the issue of biofuel and renewable energy sustainability.

ISBN: 978-1-78466-034-5 e-ISBN: 978-1-78466-036-9
Published 2015 / 256pp / £105.00

www.ingramcontent.com/pod-product-compliance
Ingram Content Group UK Ltd.
Pitfield, Milton Keynes, MK11 3LW, UK
UKHW022210051026
15624UKWH00001B/101

9 781784 660628